AF606478

ATLAS OF SCANNING ELECTRON MICROSCOPY IN MEDICINE

TSUNEO FUJITA, M. D.

Professor of Anatomy,
Niigata University School of Medicine,
Niigata, Japan

JUNICHI TOKUNAGA, M. D.

Associate Professor, Department of Microbiology,
Kyushu Dental College,
Kitakyushu, Japan

HAJIME INOUE, M. D.

Departments of Anatomy and Orthopedics,
Okayama University Medical School,
Okayama, Japan

1971

IGAKU SHOIN LTD. Tokyo

ELSEVIER PUBLISHING COMPANY Amsterdam · London · New York

Sole distributor outside Japan

ELSEVIER PUBLISHING COMPANY
335 Jan van Galenstraat
P. O. Box 211, Amsterdam, The Netherlands

ELSEVIER PUBLISHING CO. LTD.
Barking, Essex, England

AMERICAN ELSEVIER PUBLISHING COMPANY, INC.
52 Vanderbilt Avenue
New York, New York 10017

ISBN 0 444 40908 4

Printed and Bound in Japan

PREFACE

Since the seventeenth century the magnifying glass has helped our observation of minute objects. In modern laboratories a binocular magnifying glass containing complex sets of lenses is used.

These optical instruments can extend the resolution of our naked eyes only to a certain extent. Though such structures as intestinal villi and renal glomeruli can be clearly observed, we come to the point where an identification of the individual cells forming these structures is desired—and much more. It was formerly but a scientist's dream to observe the delicate processes of those cells in their three-dimensional entity. The wall we meet is based on the limitations in the depth of focus as well as in the resolution, both limitations belonging to the intrinsic nature of light. The fine structures of cellular and subcellular levels have been studied almost exclusively by cutting the tissues into very thin sections and observing them under the transmission light or electron microscope. A large gap thus has occurred between macroscopic anatomy aided by magnifying glasses and microscopic anatomy based on the findings in the sections. This gap is also that between the three-dimensional and the two-dimensional fields.

The latest advance in electronics has shed light on this problem. The appearance of the scanning electron microscope, which is based on a mechanism fundamentally different from that in the previous transmission electron microscope, has realized that dream of ours to continuously raise the magnification of the magnifying glass to the level which the electron microscope achieves.

Observing things as they are with the naked eye is without doubt the most realistic and convincing method of study. The scanning electron microscope, though mediated by the action of electrons, may be said to extend simple and direct reality or, better to say, the feeling of attendance in the macroscopic observation to the microscopic fields. It may not be exaggerating to say that the wall we believed could never be overcome has been broken down by the advent of the scanning electron microscope.

In the last three years we have attempted to observe under the scanning electron microscope a variety of materials related to biomedical fields of study. Minute representatives of Nature, one after another, revealed their fascinating figures under this electron microscope. The octopus cells extended their tentacles to the renal glomerulus, the aspergillus fungi raised their gorgeous bouquets of conidiophores, the worms of isolated mitochondria appeared to be bending and stretching their bodies... Observation of one object drove us to the study of another.

Thus the photomicrographs accumulated to a considerable number and a selection of them is now to be bound in an atlas. They were not gathered so purposefully as to be arranged under a strict system, but may be considered a miscellaneous collection. This shortcoming of our atlas, we hope, will be rather an advantage in introducing the performance of our new weapon, the scanning electron microscope, and in showing its wide applicability to biomedical studies.

The main purpose of this volume, which includes pictures of those materials whose three-dimensional fine structures were first revealed to our research group, is to stimulate new investigations in related fields. This volume may, however, also serve in the visual education of biomedical students and, further, perhaps amuse general readers as an "Album of Nature." These multiple characteristics of the scanning electron micrographs are believed not to decrease their purely scientific value but rather to indicate their many-sided merits.

On the publication of this atlas, we express our cordial thanks to those who kindly contributed their precious photomicrographs. Dr. Kéiichi Tanaka of the Tottori University Medical School provided the pictures of chromosomes and lens fibers; Dr. Yoichi Ishii of the Kyushu University School of Medicine, the figures of cercaria; Dr. Naoya Kosaka of the Okayama University Medical School, the figures of the inner ear; Dr. Takuo Ogata of the same school, the surface views of the stomach; and Dr. Takuo Jozaki of the Yahata-Seitetsu Hospital, the picture of bacteria on the tongue.

Niigata, Kokura, Okayama,
July, 1970

Tsuneo Fujita
Junichi Tokunaga
Hajime Inoue

CONTENTS

CELL ORGANELLES

GALL AND URINARY STONES

DENTAL TISSUES

PARASITES

FUNGI AND BACTERIA

What Is the Scanning Electron Microscope?

Although theoretical and experimental studies on the scanning electron microscope had already been initiated in Germany in the latter half of the nineteen-thirties, the industrial production of the scanning electron microscope was not successful until the early sixties in England and a little later in Japan.

In the scanning electron microscope an extremely narrowed beam of electrons is directed at the specimen and scanned to and fro across its surface. The secondary emissions emerging from the specimen are collected by a detector and the changes in their amount are transformed into an electric signal. This signal then is amplified and expressed in terms of a variation in brightness on the Braun tube. A picture is thus produced on a screen and can be recorded by indirect photography.

The pictures of the scanning electron microscope, as far as an even secondary emission is ensured, well represent the cubic appearance of the material surface and give an effect of illumination as if the light were put at the place of the detector and the eye of the observer at the site of the electron gun.

Materials such as metals which, when bombarded by electrons, emit ample secondary electrons can be observed under the scanning electron microscope without any treatment. However, as biological materials generally issue only scanty secondary emissions, they must be coated with a thin metallic layer beforehand. The practical procedure of coating will be described later.

Fig. 1. An example of stereo-pair micrographs. Otoconia from the macula sacculi of the rabbit. ×1,500

The resolution of the scanning electron microscope is about 250 Å. It depends naturally on the technique of sharpening the electron beam. However sharpened it may be, the secondary emission emerges not from the very point hit by the electron beam but from a certain area and depth. This nature of electrons significantly affects the resolution of the scanning electron microscope. Furthermore, the coating itself, when it is made, substantially decreases the resolution. Because of these conditions, the highest effective magnification of the scanning electron microscope at present is about 30,000 times. This magnification can be and should be secured by careful maintenance and adjustment of the microscope, though many researchers today seem to be satisfied with a magnification of several thousand times.

One of the great merits of the scanning electron microscope is the conspicuously large depth of focus. Simply speaking, the depth of a field roughly corresponds to half of its width. When we have a field of 1 mm in diameter, a depth of 0.5 mm is sharply focused. The scanning electron micrographs are thus characterized by the sharp appearance of every elevation and depression on the surface insofar as the peak-to-valley height does not exceed the above-mentioned range. If we remind ourselves of the irritating experiences in close-up photography and light microphotography that are focused only to an extremely limited depth, the scanning electron microscope can be said to have a revolutionary performance.

By tilting the specimen holder through an adequate angle between two exposures, stereo-pair micrographs are obtained, which, when observed through a stereo-viewer, afford a real three-dimensional view of the specimen surface (*Fig. 1*). One may also measure the peak-to-valley height of given objects on the surface using a stereometer.

Scanning electron microscope observation of biological materials was retarded in comparison to that of hard materials such as metals and crystals, and it was only in 1967 that systematic studies began to be published. The hard tissues, such as teeth and bones, are relatively simple to treat, but in soft tissues, cells, parasites and microbes, it has been a large problem as to how to minimize the deformation caused by drying. The technique of metal coating should also be improved in order to clarify the fine structures of subcellular level.

Methodology in the sample preparation, including these problems, will now be briefly described.

Preparation of Specimens

The object of scanning electron microscopy is the *surface* of materials. The surfaces, however, are not necessarily natural and free ones, but may be artificial ones made by cutting, tearing and even by crushing. Free cells, subcellular fractions, as well as formed elements and crystals in body fluids and drugs are appropriate materials for scanning electron microscopy. Various kinds of impressions, such as simple celluloid replicas of surface reliefs and plastic casts of cavities and vessels, also provide useful specimens (Fujita, Tokunaga and Inoue, 1969).

The potential materials for scanning electron microscopy thus may be said to be of infinite variety. This also implies a possibly large variety of methods in specimen preparation, and it should be understood that the following comment on the specimen preparation of animal tissues and free cells is just one possible choice.

1. Fixation

When the surface to be observed is covered by mucous and other substances, attempts should be made to eliminate them as completely as possible. Use of surface-activating materials or enzymes may be effective, but a simple and unexpectedly effective method is patiently rinsing the surface with a jet-stream of Ringer's solution or physiological saline.

As fixative, 2.5 per cent glutaraldehyde (in a 0.1 M phosphate buffer pH 7.2) is recommended for general use. A comparably good result for some materials may be obtained by simple fixation in 10 per cent formalin.

These non-coagulant fixatives have the merit of preventing proteinic fluids covering the surface from forming coagulations which would hide the surface appearance. On the other hand, however, the tissues and cells may be insufficiently hardened by these fixatives so that they may undergo severe shrinkage and deformation during the later procedures of dehydration and drying. This deficiency in aldehyde fixatives may be overcome by a post-fixation in 2 per cent osmium tetroxide, 0.2 per cent corrosive sublimate or in 2 per cent potassium permanganate (*Fig. 2*).

Though both glutaraldehyde and formalin can be applied to free cells, our experience with red blood corpuscles, which are especially sensitive to osmotic pressure, indicates that a glutaraldehyde concentration of 1 per cent (in a 0.1 M phosphate buffer) preserves the most natural forms of cells (Tokunaga, Fujita and Hattori, 1969).

In the case of blood corpuscles and spermatozoa, it is of primary importance that the blood or sperm plasma be eliminated so that the cellular elements become virtually naked. Rinsing in physiological saline is effective, but a better and more simple method is to drop a small amount (2 or 3 drops) of blood or semen into a large volume (20 ml) of a non-coagulant fixative and to shake vigorously. The cellular elements are thus extensively dispersed in the fixative and, if one removes the supernatant after gentle centrifugation, the plasma which when coagulated would cover and bury the cells as "mud" under the scanning electron microscope, can be eliminated virtually completely (Fujita, Miyoshi and Tokunaga, 1970).

Specimens of subcellular fractions can be prepared by essentially the same method as described above (KURAHASI et al., 1969).

After the fixation the tissues and cells are rinsed and then dehydrated.

2. Dehydration and Drying

If fixed or unfixed tissues in a wet state are dried in air, a severe shrinkage and deformation is inevitable. In the case of free cells, drying water, with its high surface tension, gathers all the muddy substances from the vicinity to the surface of the cells and one may see under the scanning electron microscope only cells buried in the "mud." To minimize these changes, the following method of acetone dehydration and air drying is recommended.

The tissues and cells are dehydrated in acetone of ascending concentrations, i.e., in 30, 50, 70, 90, 95 and 100 per cent. The time for every concentration is probably 10 minutes, though naturally it depends on the size of the specimens. The final stage of dehydration should be done carefully through several changes of 100 per cent acetone and dried acetone. The specimens are then put on a filter paper and dried in air.

The method of acetone dehydration and air drying was reported by BARBER and BOYDE (1968). A little later, but independently, we came to know the usefulness of volatile media in drying tissues. Ethanol, acetone and propylen oxide were all found effective, but we found that moderately volatile acetone was most easy to handle, thus reaching the same conclusion as BARBER and BOYDE (FUJITA, INOUE and KODAMA, 1968).

Since 1969, however, BARBER and BOYDE have "changed more or less exclusively to freeze-drying of soft tissue specimens because of the great degree of shrinkage that can occur on air-drying." Drops of water containing either fixed or unfixed specimens were frozen in a bath of dichlorodifluoromethane cooled by liquid nitrogen. After being transferred to liquid nitrogen, they were put on a cold probe to be dried in vacuo.

We have experienced that freeze-drying may also cause relatively marked artifacts on one hand, and on the other hand, deformations from the acetone-drying method may be prevented to a considerable extent by various devices in the procedures of post-fixation and drying (*Fig. 2*). We attach importance to making the most of the merits of simple and rapid specimen preparation in scanning electron microscopy. From this standpoint we have dared to stay with the acetone drying and almost all the pictures in this volume are based on this method.

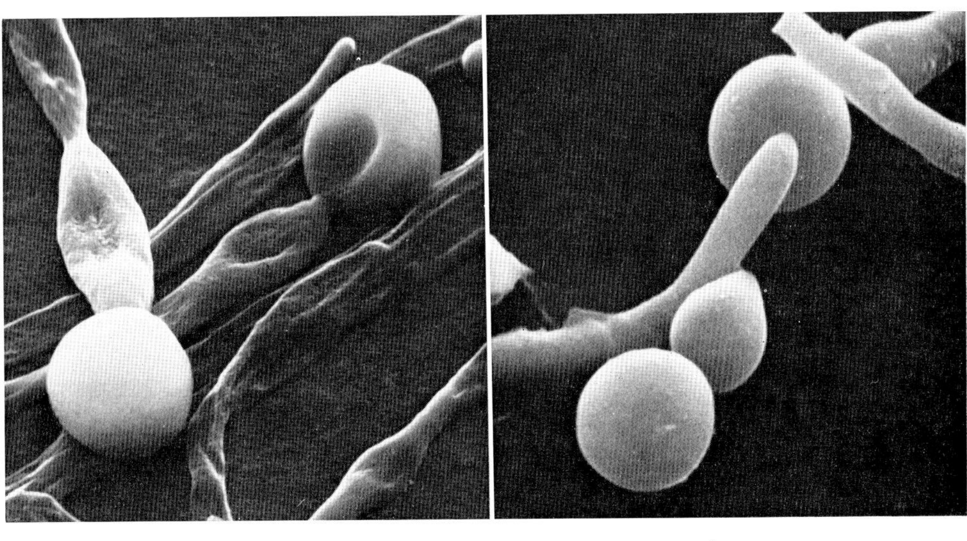

A B

Fig. 2. Scanning electron micrographs of *Candida albicans*.
A: Fixation in 2 per cent glutaraldehyde, acetone dehydration and air drying.
B: Postfixation in 2 per cent $KMnO_4$ was made after the glutaraldehyde fixation. The collapse and shrinkage of the fungal tissue which are obvious in Fig. A have been prevented by the postfixation as shown in Fig. B. ×3,250

Dehydration and drying of free cells (of blood, semen, ascites, etc.) and subcellular fractions are performed by the following procedures.

The fixed specimens are advanced, gentle centrifugation being repeated, through ascending grades of acetone. The final stage of dehydration in dried acetone should be done especially carefully. By vigorously shaking the centrifuge tube the specimen is emulsified in a few ml of dried acetone. This emulsion is dropped with a pipet on small pieces (1 × 1 cm) of glass slide and dried under a warm breeze from a hair-dryer.

We recommend that the dried specimens be carefully checked before the next procedure of metal coating. In the case of tissues, the surfaces to be used should be surveyed under a binocular to see whether the structures of interest are free of coverage and what artificial deformations have occurred. In the case of free cells, specimens with fewer deformities and better scattering of cells covered by no muddy substance are selected under a transmission light microscope. Even such minute particles as blood platelets and isolated mitochondria can be clearly examined by this method. Dark field illumination is often useful for these materials.

3. Metal Coating

The next procedure is coating the specimens with carbon and metals. This procedure has two purposes. One is to ensure the electric conductivity from the specimen surface to the holder and another is to make a layer which will emit ample secondary electrons upon bombardment by electrons.

Though a metal coating was once made by simply dipping the specimens into a saturated ethanolic solution of auric chloride and drying them in air (JAQUES, COALSON and ZERVINS, 1965), the prevailing method now is vacuum evaporation.

Double coating with carbon and gold by vacuum evaporation (BARBER and BOYDE, 1968) seems to be most widely performed. Carbon assures a tough layer of a high electric conductivity and is said to be evaporated going not directly but "around corners." Gold is easy to treat in evaporation, emits ample secondary electrons and is free from degeneration. The specimens shown in this volume are, for the major part, doubly coated with carbon and gold, but a part of them are singly coated with gold. It is difficult to come to a definite conclusion but we are inclined to find no significant difference between the simple and double coatings. Platinum-palladium, silver, copper or aluminum may be used for gold.

It is of critical importance for successful scanning electron microscopy to have a coating layer cover the specimen as evenly as possible in spite of the roughness in its surface. This purpose is attained to some extent either by setting two or three evaporation sources above the specimens or by repeating evaporation from different directions.

Much more effective is the use of a specimen disk which can be rotated around the vertical axis (BARBER and BOYDE, 1968). The evaporation source is set obliquely above the disk (at an angle of 40–60°) and the disk, during the deposition of carbon and metals, is rotated quickly by hand or, still better, by motor drive. Recently a more advanced evaporator has been available in which the specimen disk is motor driven into a combined movement of a horizontal rotation around the vertical axis and a rocking (to a desired extent of angle) around the horizontal axis. This device guarantees much more even coating than in the case of simple rotation. Carbon is evaporated in a vacuum of about 5×10^{-5} Torr, while gold in a vacuum a little lower than this.

According to the combined scanning and transmission electron microscope study by BARBER and BOYDE (1968), an appropriate coating layer consists of 200 Å thick carbon and 300 Å thick gold in the case of double coating of soft tissue surface. Some authors, however, believe that the thickness of the coating should be as thin as 100 Å.

Recently an evaporator equipped with a "thickness monitor" has been provided which, by the use of a quartz oscillator, enables the detection of ultra-fine weight changes caused by metal deposition. With this instrument one can attain an adjustable and repeatable deposition of coating layers.

When the coating is too insufficient to ensure a good electric conductivity on the specimen surface, the image produced is severely disturbed by an undesirable charging effect. When the coating is superfluous, the surface structures to be observed are covered "under the snow."

4. Electron Microscopy

The specimens are, either before or after metal coating, mounted on a copper stub with conductive silver-paste and inserted into the specimen chamber of the scanning electron microscope. A beam accelerating voltage is selected between 5 and 25 kV. In a low magnification, a high contrast in images free of charging effect can be obtained at 5 or 9 kV. The highest accelerating voltage, 25 kV, is needed only for magnifications higher than 20,000 times.

When one compares the images of the same specimen under the same magnification at different accelerating voltages, a significant difference may be noticed. A lower voltage generally gives a sharper picture of the surface structure. The difference is ascribed to the fact that at lower voltage only the secondary emissions from the very superficial layer of the specimen are collected, whereas at higher voltage, information from deeper layers becomes mingled.

Though the scanning electron micrographs have the merit of being very simply understood, the following points may be worthy of attention:

First, the scanning electron micrographs are, so to speak, "snow-covered landscapes." The surface structures appear thicker because of the depth of the snow or coating layer. There is no significant problem in structures larger than cellular level but minute matters of subcellular level should be carefully analyzed. Projections appear thicker and depressions smaller than they really are. Very narrow spaces may be buried.

Second, the more or less marked shrinkage of the specimens during the drying procedure should be taken in account. A comparison of the dimensions of structures in scanning and transmission electron micrographs thus requires careful consideration.

Reference

Barber, V.C. and A. Boyde: Scanning electron microscopic studies of cilia. Z. Zellforsch. 84: 269-284 (1968).

Boyde, A. and V.C. Barber: Freeze-drying methods for the scanning electron-microscopical study of the protozoon *Spirostomum ambiguum* and the statocyst of the cephalopod mollusc *Loligo vulgaris*. J. Cell Science 4: 223-239 (1969).

Boyde, A. and C. Wood: Preparation of animal tissues for surface-scanning electron microscopy. J. Microsc. 90: 221-249 (1969).

Cosslett, V.E.: Scanning microscopy with electrons and X-rays. J. Electron Microsc. 16: 51-64 (1967).

Fujita, T., H. Inoue and T. Kodama: Scanning electron microscopy of the normal and rheumatoid synovial membranes. Arch. histol. jap. 29: 511-522 (1968).

Fujita, T., J. Tokunaga and H. Inoue: Scanning electron microscopy of the skin using celluloid impressions. Arch. histol. jap. 30: 321-326 (1969).

Fujita, T., M. Miyoshi and J. Tokunaga: Scanning and transmission electron microscopy of human ejaculate spermatozoa with special reference to their abnormal forms. Z. Zellforsch. 105: 483-497 (1970).

Jaques, W.E., J. Coalson and A. Zervins: Application of the scanning electron microscope to human tissues. A preliminary study. Exp. molecul. Pathol. 4: 576-580 (1965).

Kurahasi, K., J. Tokunaga, T. Fujita and M. Miyahara: Scanning electron microscopy of isolated mitochondria. I. Arch. histol. jap. 30: 217-232 (1969).

Tokunaga, J., T. Fujita and A. Hattori: Scanning electron microscopy of normal and pathological human erythrocytes. Arch. histol. jap. 31: 21-35 (1969).

Tokunaga, J., T. Fujita and H. Inoue: Medical application of scanning electron microscopy (in Japanese). Igaku no Ayumi 68: 485-491 (1969).

Scanning Electron Microscopy 1968. Proceeding of the Symposium on "The Scanning Electron Microscope — the Instrument and Its Applications" (Symposium Director: O. Johari). IIT Research Institute, Chicago, 1968 (185 pages).

Scanning Electron Microscopy 1969. Proceeding of the Second Annual Scanning Electron Microscope Symposium (Symposium Director: O. Johari). IIT Research Institute, Chicago, 1969 (525 pages).

Scanning Electron Microscopy 1970. Proceeding of the Third Annual Scanning Electron Microscope Symposium (Symposium Director: O. Johari). IIT Research Institute, Chicago, 1970 (534 pages).

VISCERAL ORGANS

1. Filiform Papillae of the Tongue

The back of the tongue feels rough as it is covered with tiny spine-like protrusions armed with cornified epithelial cells. These protrusions are named filiform papillae to be differentiated from the larger and less numerous fungiform, folliate and vallate papillae. In contrast to the latter forms, the filiform papillae lack taste buds and serve simply as a mechanical file in licking off things. As slight movement and pressure on

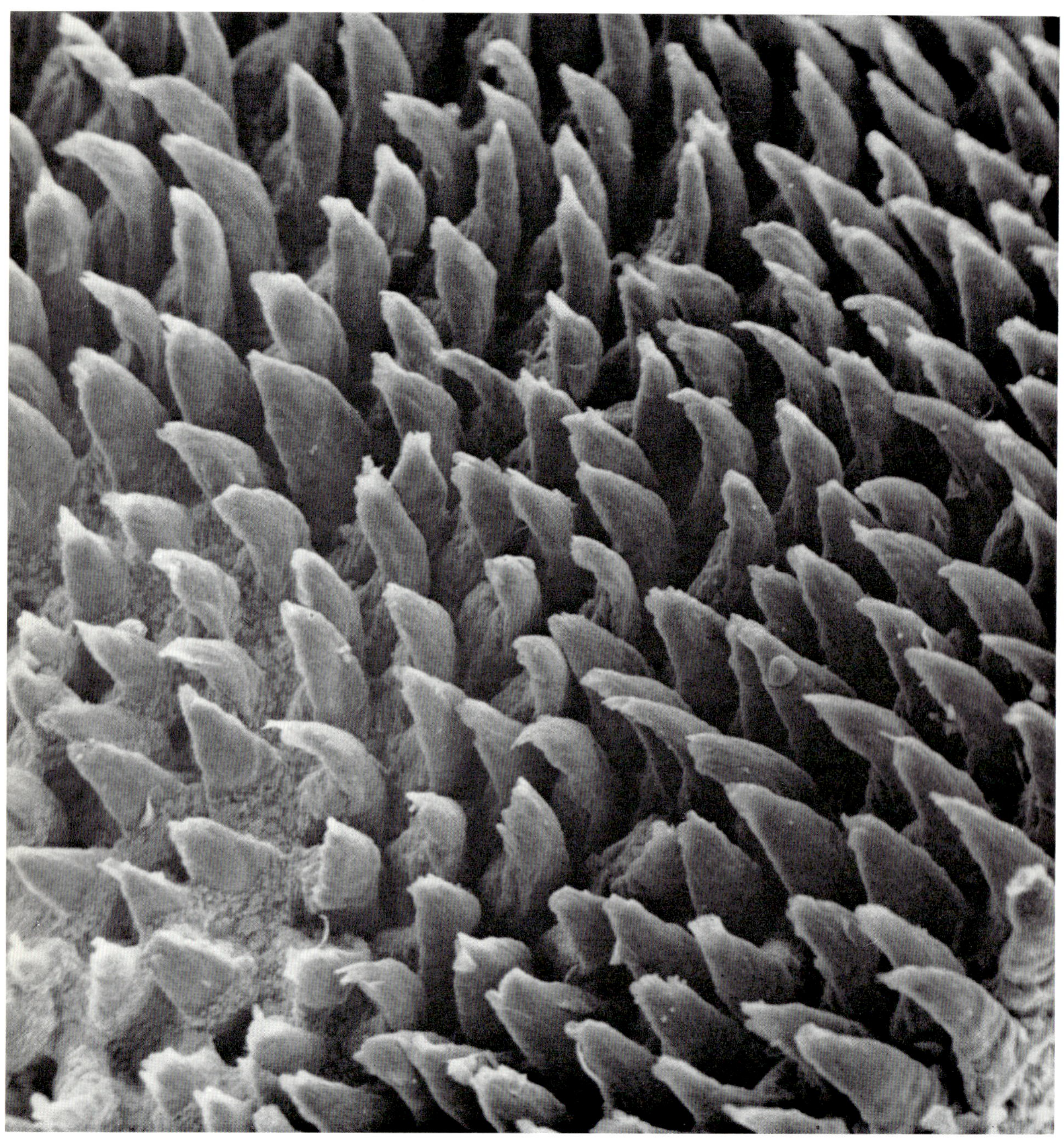

Fig. 3 (× 170)

Fig. 4 (×430)

their ends may be transmitted to the nerve endings richly distributed at their base, it has been presumed that they are also a kind of tentacle for the perception of touch sensation.

Figs. 3 and **4** show the filiform papillae densely covering the back of the tongue of a rabbit. Every papilla is curled with its end pointed to the throat, a form suited for the licking-off mechanism of the tongue.

Fig. 5 (× 1,700)

Fig. 5 which is a closer view of the middle portion of the preceding micrograph reveals scale-like epithelial cells covering the papilla. In **Figs. 6** and **7** an exquisite relief on the surface of these cornified cells is shown highly magnified. Fig. 7 includes three of these cells piled like slates.

The micrographs shown in this section clearly indicate the two merits of the scanning electron microscope: that it can demonstrate the surface structures of matters from the level of a magnifying glass to that of an electron microscope, and that it affords such a large depth of focus as could never be expected in light microscopy.

Note

The back of the rabbit tongue was carefully washed with water and fixed in 10 per cent formalin. Small blocks of the tissue were dehydrated in acetone, dried in air and coated with gold. EM: JSM-2

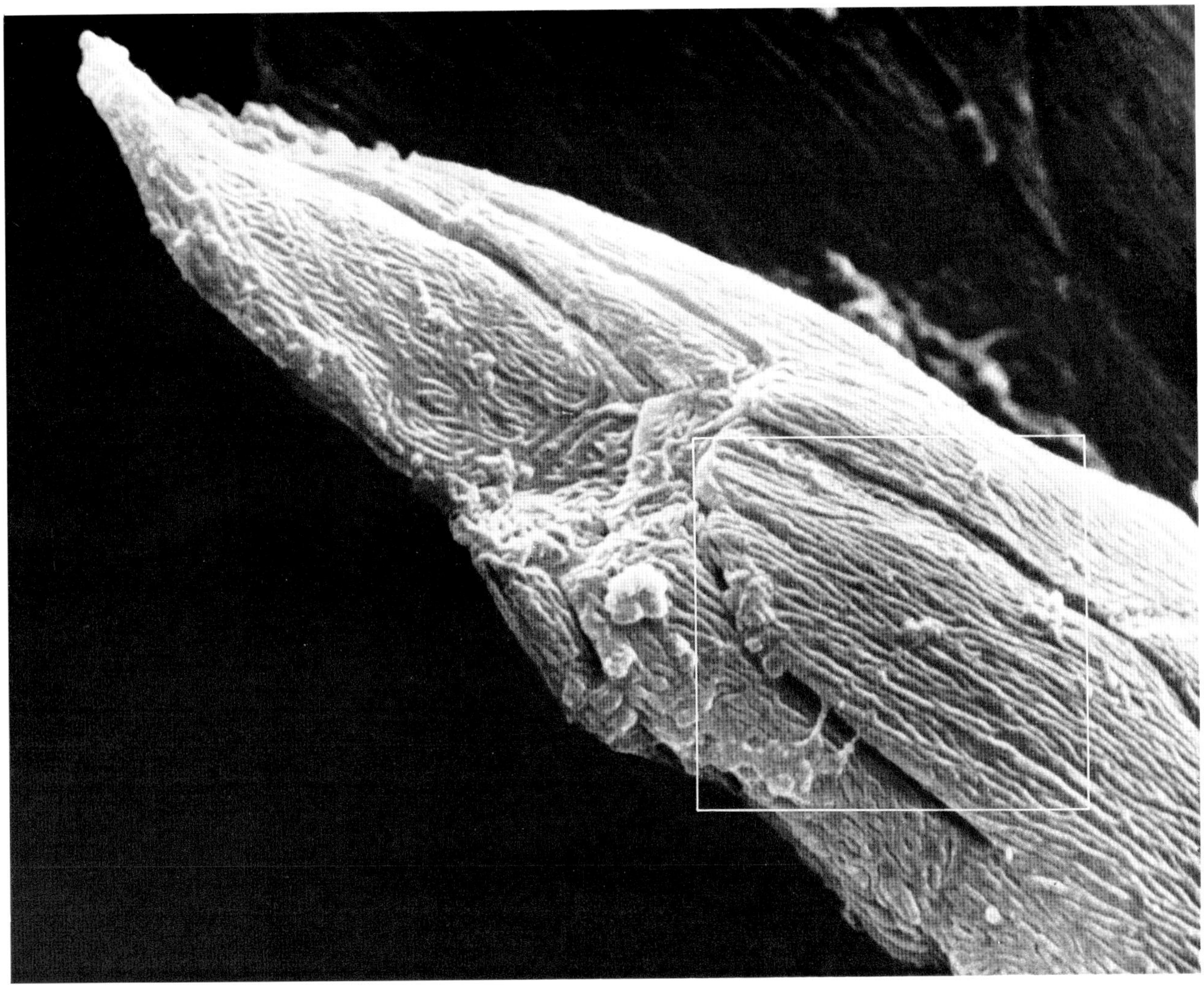

Fig. 6 (×5,100)

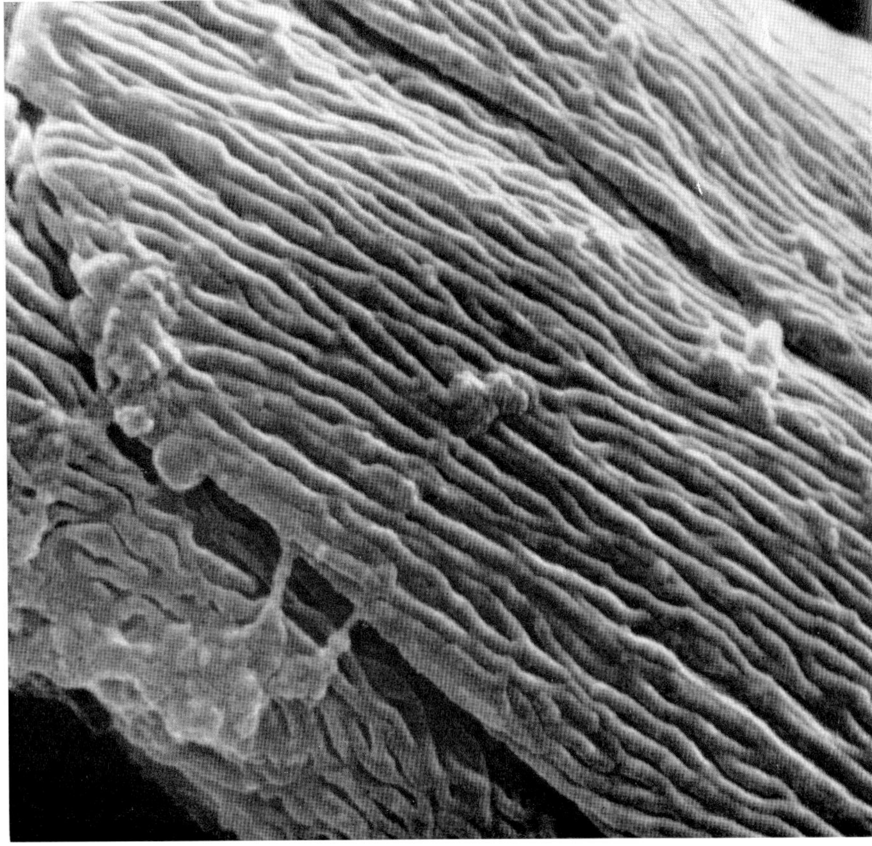

Fig. 7 (×12,000)

2. Sensory Papillae of the Frog Tongue

In the frog, which has no taste buds on the tongue, the fungiform papillae play the role of taste receptors. It is also known that these sensory papillae, dispersed on the back of the tongue, are contractile in response to various physical and chemical stimuli.

Under the scanning electron microscope a fungiform papilla looks like a round table (**Fig. 9**). The main central part of the table is called a sensory epithelial disc and is covered by a number of hexagonal sensory cells. Surrounding this epithelial disc there are a few rows of ciliated cells and the side of the table is covered by large rounded epithelial cells.

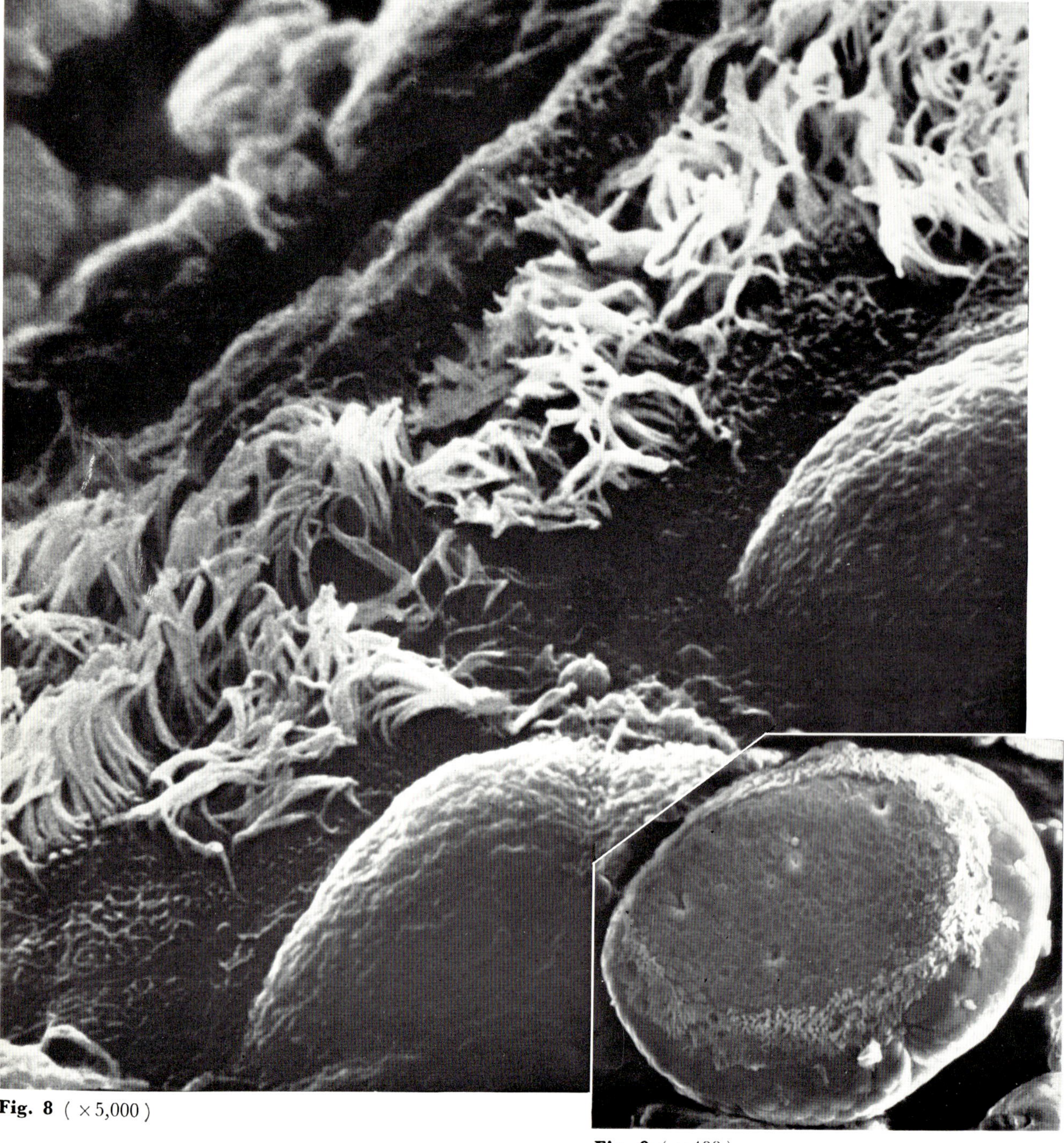

Fig. 8 (×5,000)

Fig. 9 (×400)

Fig. **10** (× 15,000)

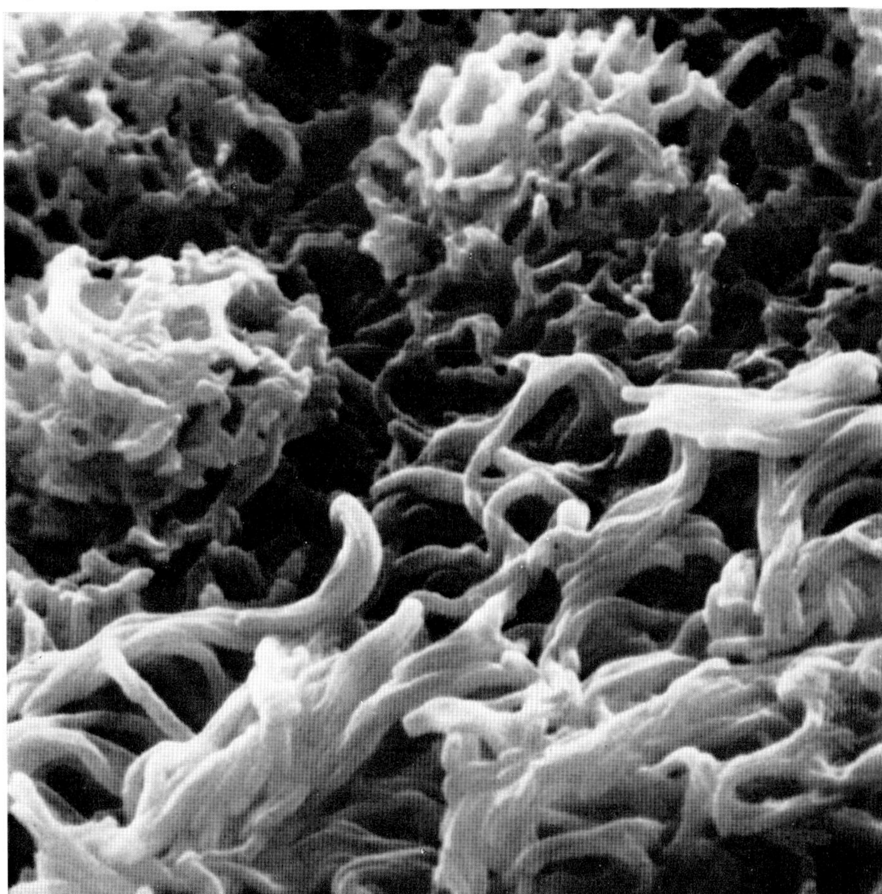

Fig. 8 is a view of the peripheral part of the table showing the latter cells in the foreground and the cilia in the back.

Fig. 10 is a close-up of the border of the sensory epithelial disc. Noodle-like cilia are seen in the lower half of the picture. In the upper half three sensory cells are recognized by their round heads covered with shaggy microvilli.

Fig. 11 shows an aberrant ciliated cell occasionally found among the epithelial cells of the table side.

These findings were obtained in collaboration with Dr. AKITATSU SHIMAMURA, Department of Oral Histology, and Dr. SATOSHI NAKAHARA, Department of Physiology of the Kyushu Dental College.

Note

The tongue of the bullfrog was removed and kept in physiological saline for several hours. The surface of the tongue meanwhile was washed repeatedly with a jet-stream of the saline in order to remove the mucus on the tongue as completely as possible. The specimen then was fixed in 2.5 per cent glutaraldehyde (0.1 M phosphate), post-fixed in 2 per cent osmium tetroxide, dehydrated in acetone, dried in air and coated with carbon and gold. EM: JSM-2

Reference

SHIMAMURA, A. and J. TOKUNAGA: Scanning electron microscopy of sensory (fungiform) papillae in the frog tongue. Proc. 3 rd Annu. Sympos. SEM. 1970. Chicago, IIT Res. Inst. (p. 227-232).

Figs. 9 and 11 by courtesy of the Editor of the Symposium Proceeding and the IIT Research Institute.

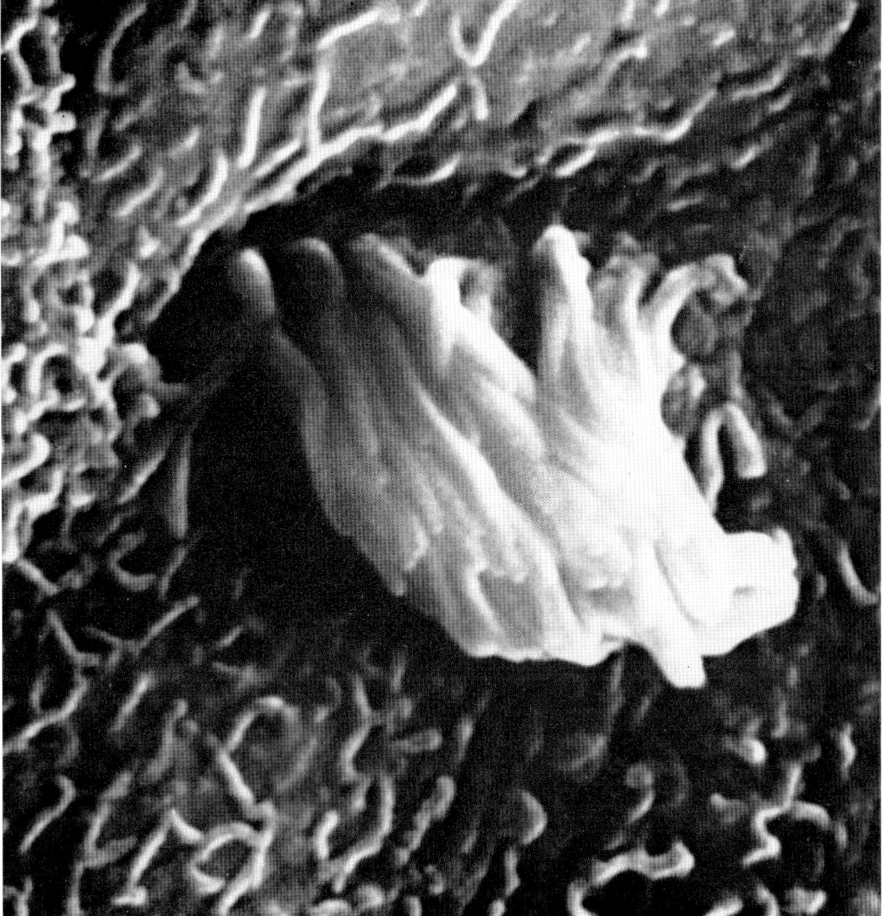

Fig. 11 (× 12,000)

3. Mucous Secretion in the Stomach

The inner surface of the stomach shows tiny, deep grooves called gastric foveolae to the bottom of which open the glands secreting hydrochloric acid and proteolytic enzyme. When we study the sections of the stomach, we are often apt to regard the foveolae as round pits. The scanning electron micrographs of the stomach surface clearly show us that the majority of the foveolae are crevasses rather than pits.

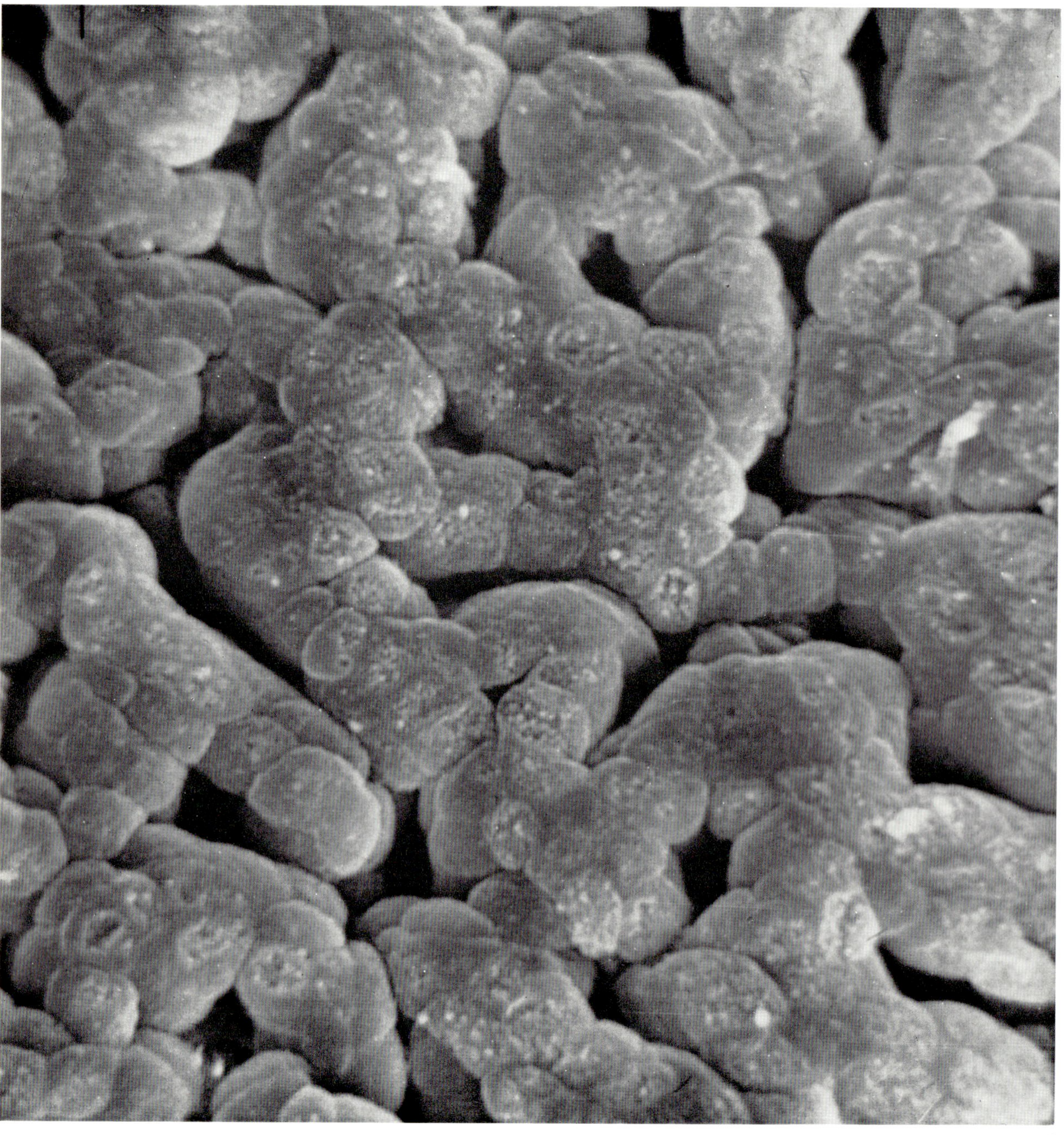

Fig. 12 (× 1,700)

Fig. 13 (× 3,600)

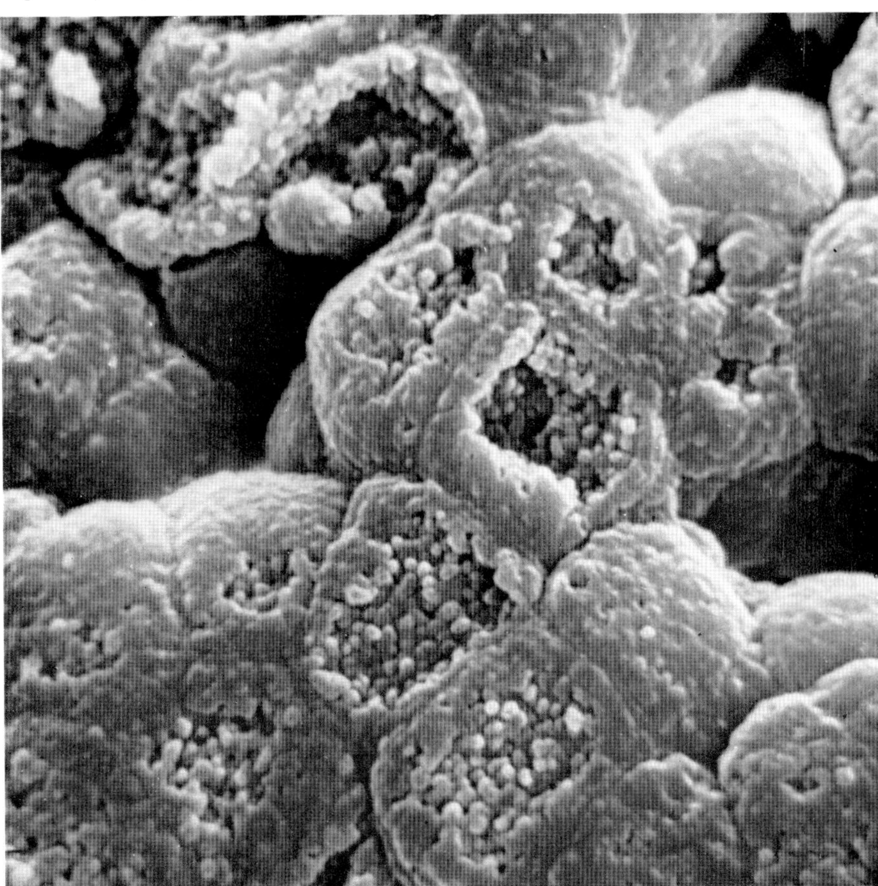

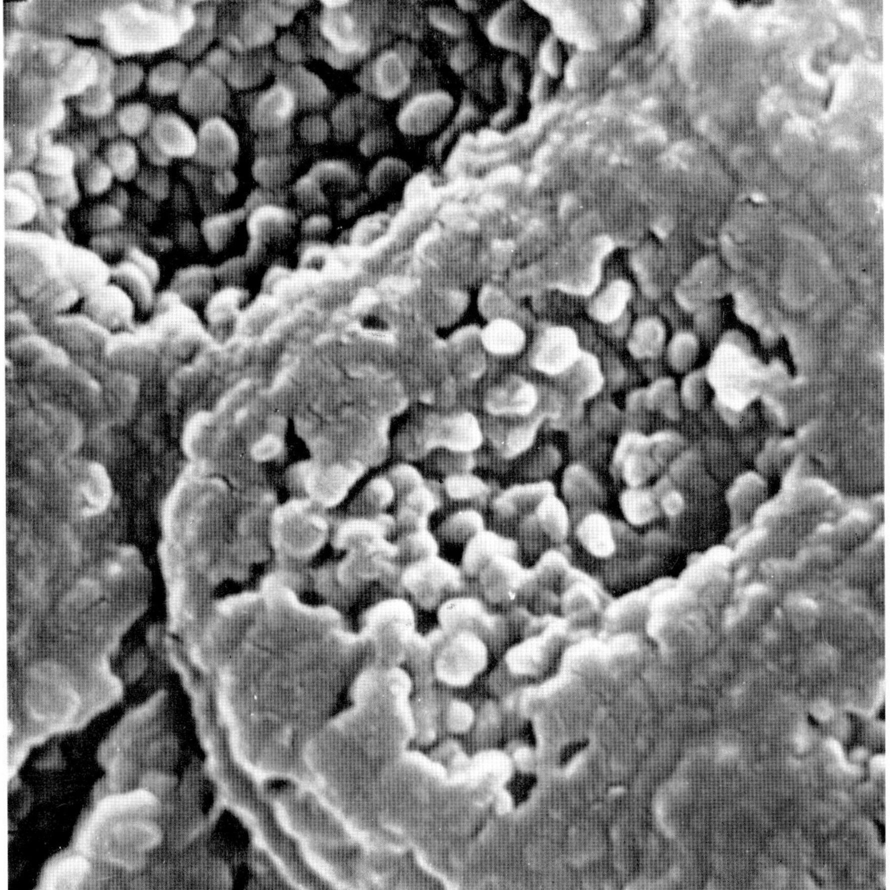

Fig. 14 (× 12,000)

Fig. 12 shows a low-power view of the rat stomach. The surface cell cords winding around the foveolae of different shapes remind us of a view of a microscopic section of the liver.

The cells covering the surface of the stomach and lining the foveolae are known to secrete a special type of mucous substance which is resistant to hydrochloric acid and protects the stomach wall against this acid. This mucous substance is known through the transmission electron microscope study of the sections of the stomach to form secretory granules of a small size.

Though already visible in Fig. 12, the surface cells releasing the secretory granules are even more clearly shown in **Figs. 13** and **14.** It seems worthy of attention that cells in different phases of secretion, from those covered with a cytoplasmic sheet to those widely exposing the granules, are dispersed in small groups.

The figures in this section were provided by courtesy of Dr. TAKUO OGATA, Department of Surgery, Okayama University Medical School and the Editor of the Tohoku J. exp. Med.

Note

Pieces of the rat stomach were fixed in 10 per cent formalin, dehydrated in acetone, dried in air and coated with gold. EM: JSM-2

Reference

OGATA, T. and F. MURATA: Scanning electron microscopic study on the rat gastric mucosa. Tohoku J. exp. Med. 99: 65-71 (1969).

4. Intestinal Villi

The inner surface of the small intestine is densely covered by mucous membrane projections about one mileimeter long. These intestinal villi not only enormously enlarge the area for the absorption of nutritious substances but also actively serve the absorptive function because their exquisite movement brings them into more effective contact with nutrients.

Fig. 15 is the surface view of a middle portion of the rabbit small intestine. The villi of a leaf-like shape resemble those in the human duodenum.

Fig. 16 shows the villi in a lower portion of the small intestine at higher magnification. The boundaries of individual epithelial cells covering the villi appear open, but this is probably an artifact occurring in the drying process of the specimen. The round, snap-like grooves dispersed here and there correspond to goblet cells which secrete a mucous substance.

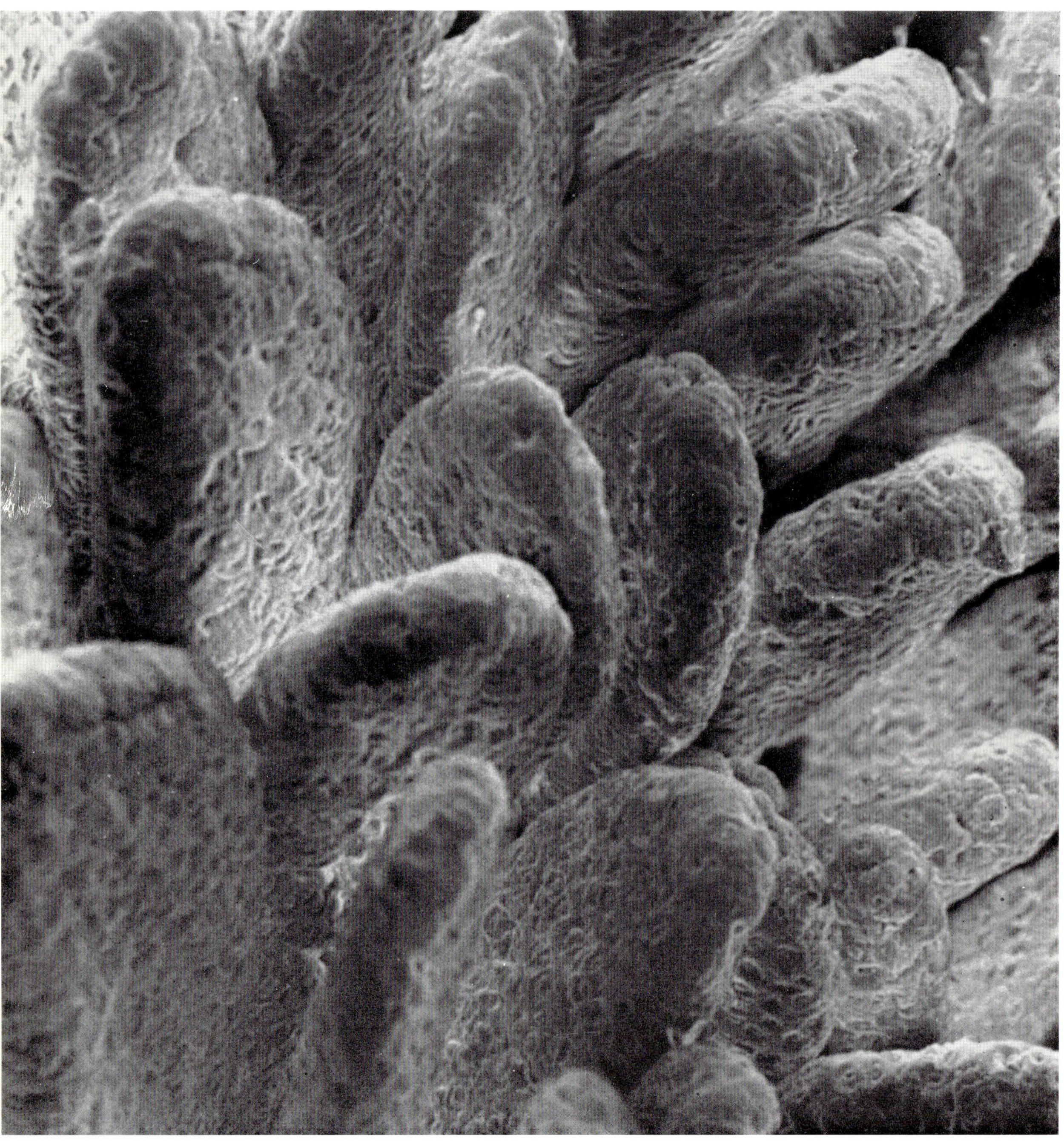

Fig. 15 (×500)

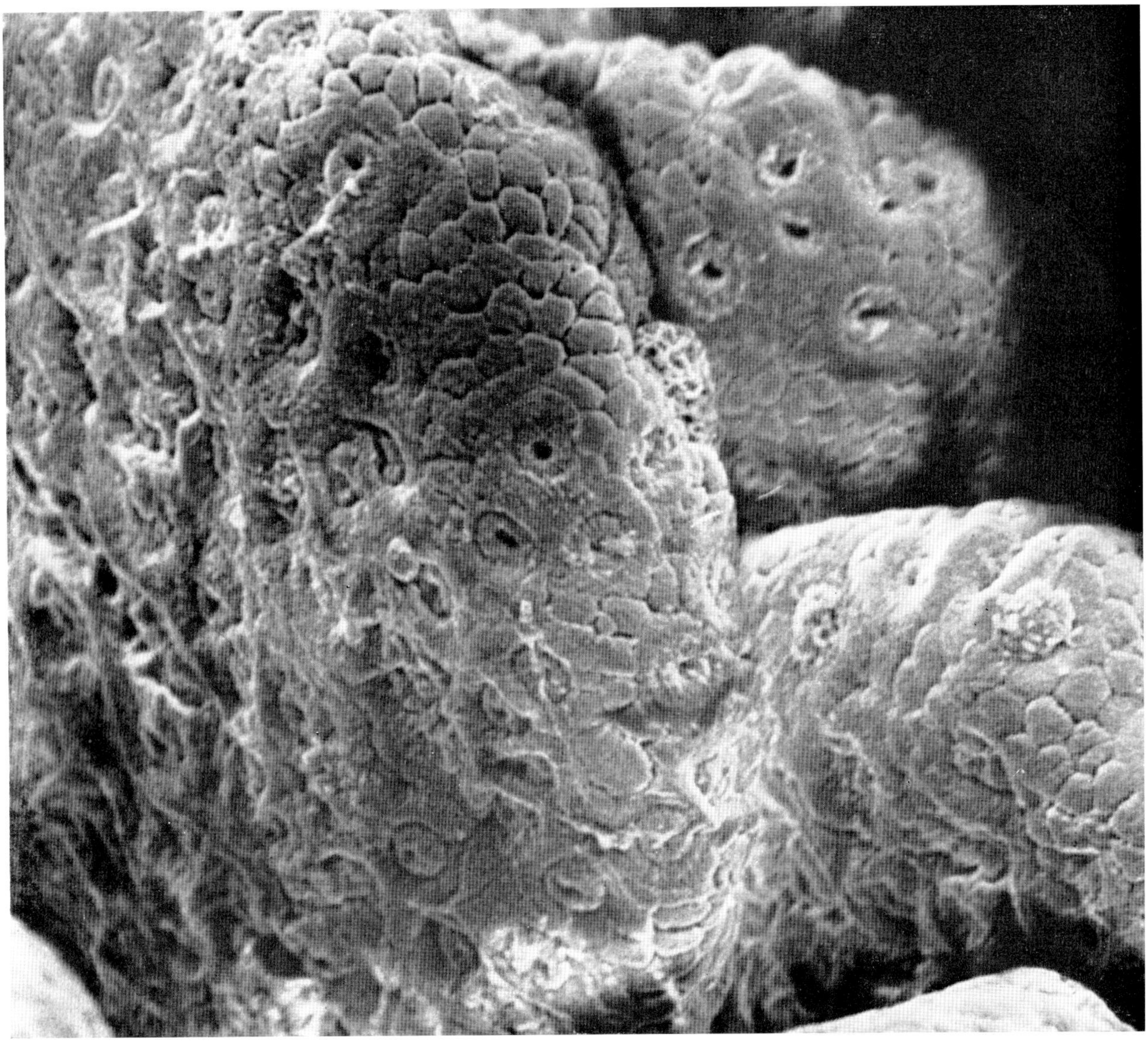

Fig. 16 (× 1,500)

Fig. 17 is a close-up of a goblet cell which extrudes its secretion in the form of a fountain.

Note

Pieces of rabbit small intestine were fixed in 2.5 per cent glutaraldehyde (0.1 M phosphate buffer), dehydrated in acetone, dried in air and coated with carbon and gold. EM: JSM-2

Reference

Marsh, M.N. and J.A. Swift: A study of the small intestinal mucosa using the scanning electron microscope. Gut 10: 940-949 (1969).

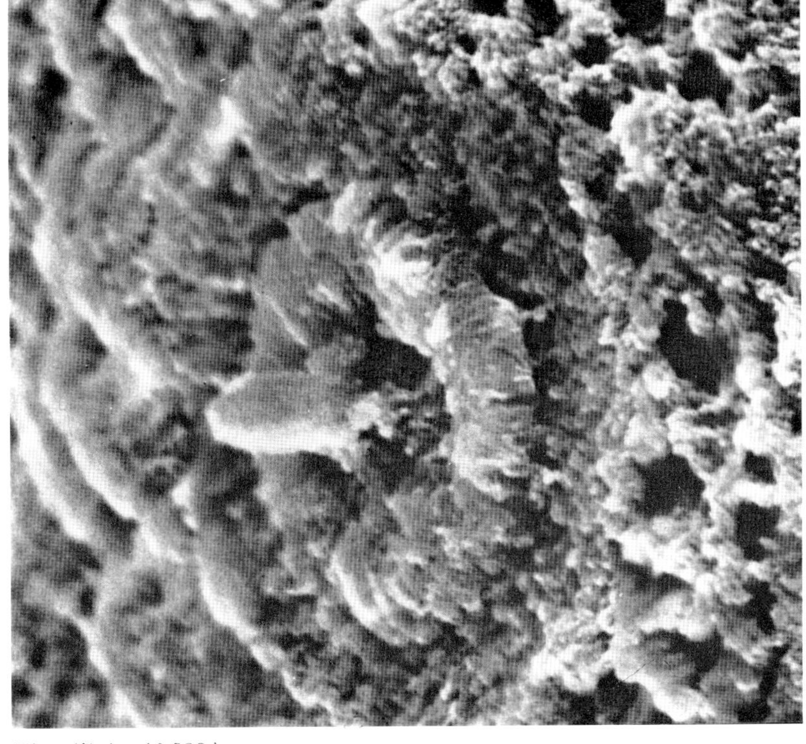

Fig. 17 (× 12,000)

5. Vermiform Appendix of the Rabbit

The vermiform appendix in the rabbit is a large lymphoid organ over 10 cm in length. Recently it has gained increasing attention as an important material for immunological studies. A naked eye observation of the inner surface of this organ reveals numerous lymph follicles arranged at regular intervals. **Fig. 18** shows this view under a low-power scanning electron micrograph. Under the domes of lymph follicles, each one surrounded by a deep groove, are hidden innumerable lymphocytes and plasma cells which are known as the source and storage of antibodies.

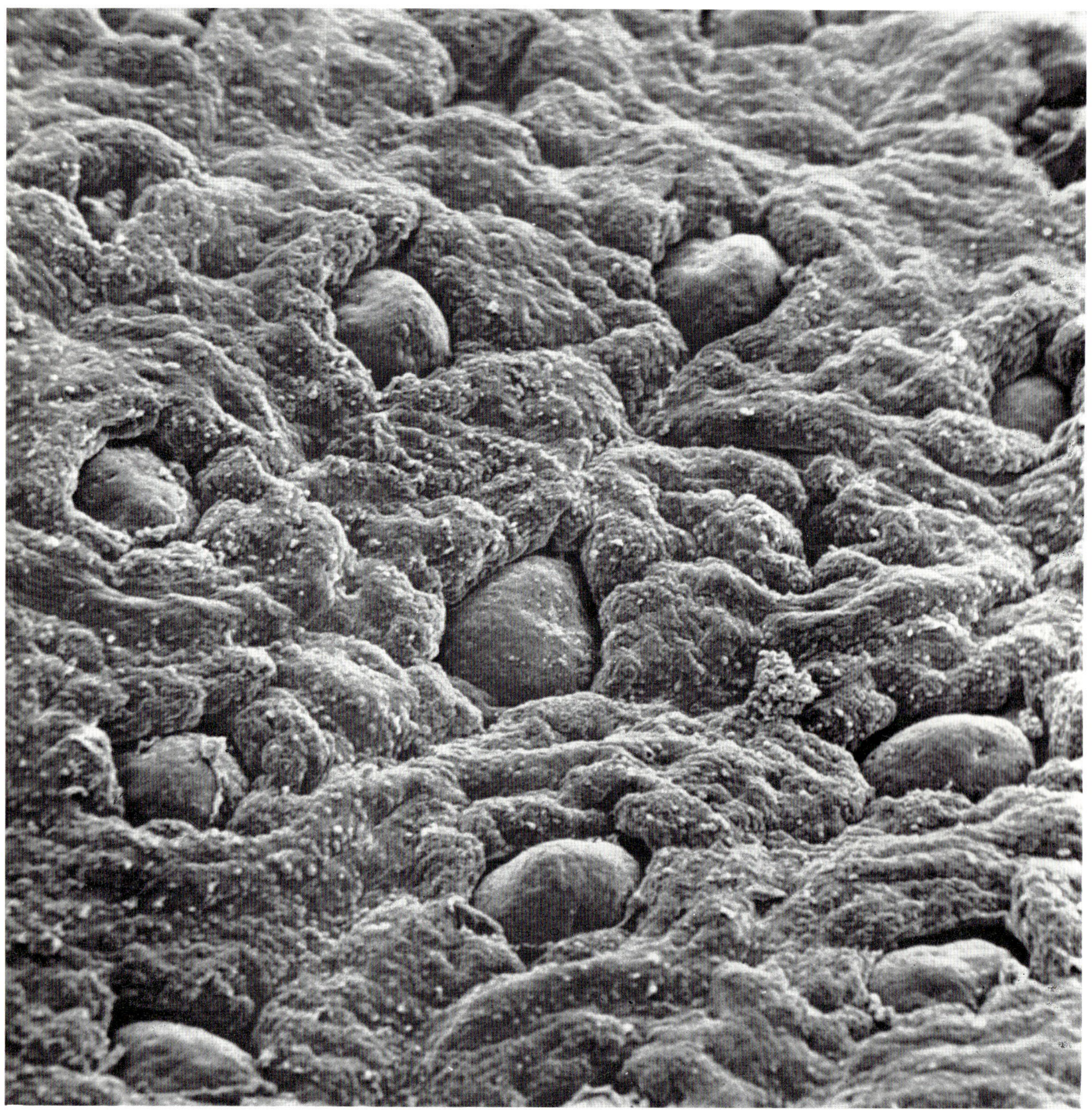

Fig. 18 (×170)

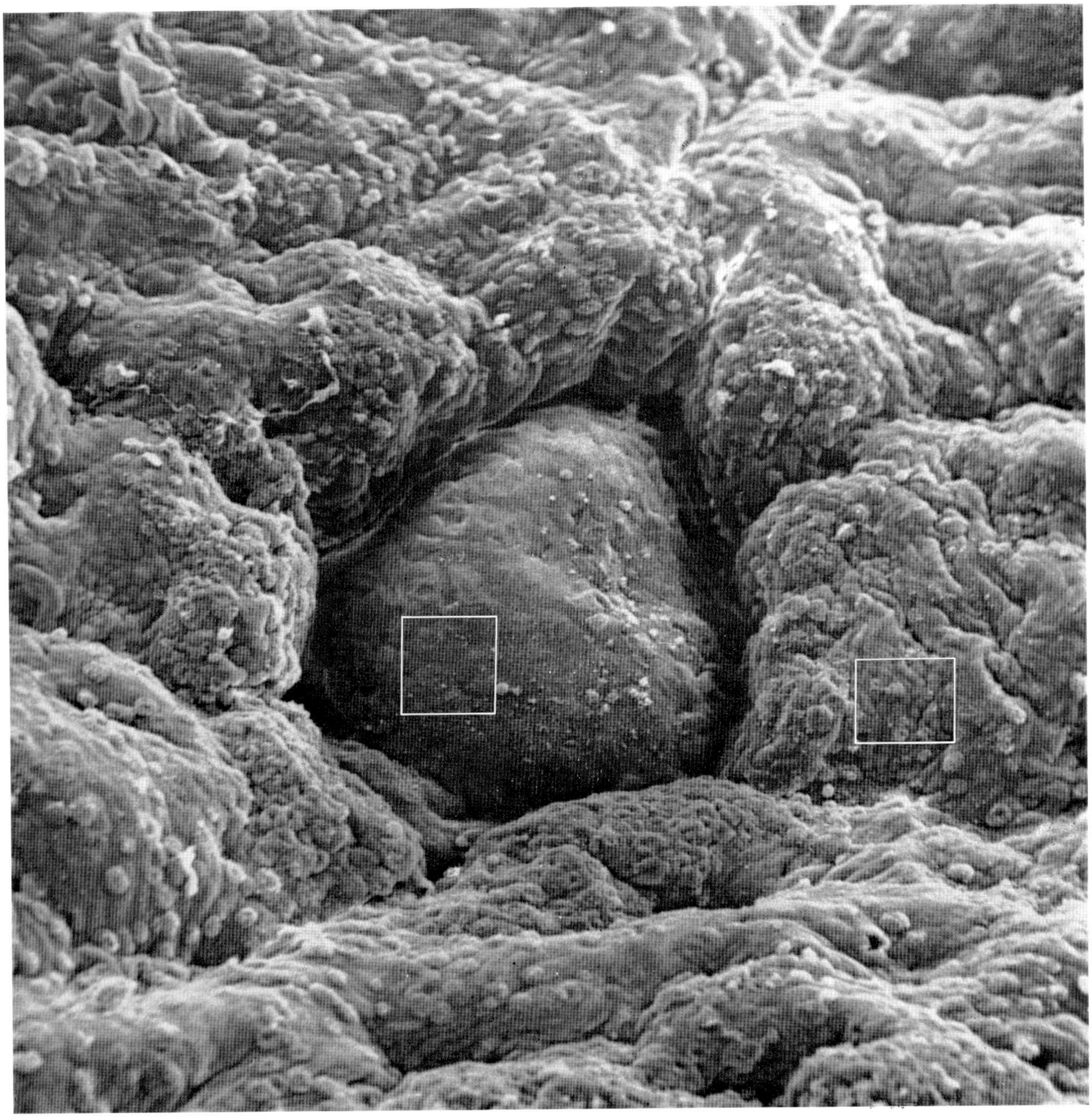

Fig. 19 (×510)

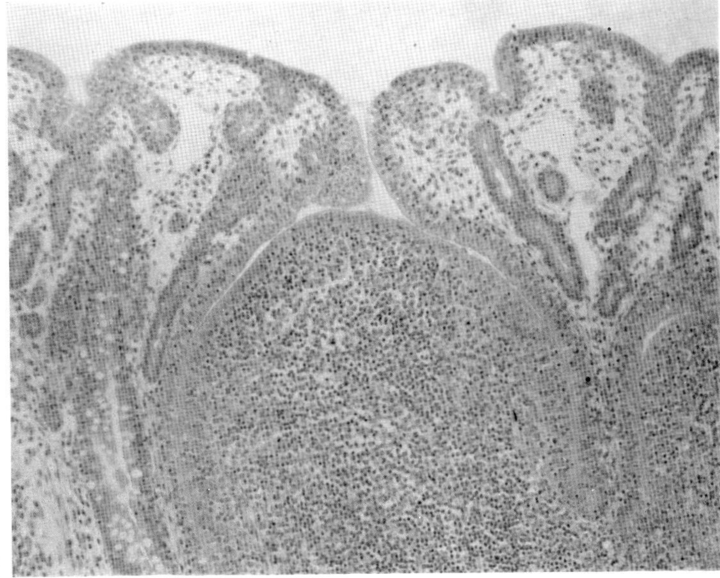

Fig. 20 (×350)

Fig. 21 (×3,600)

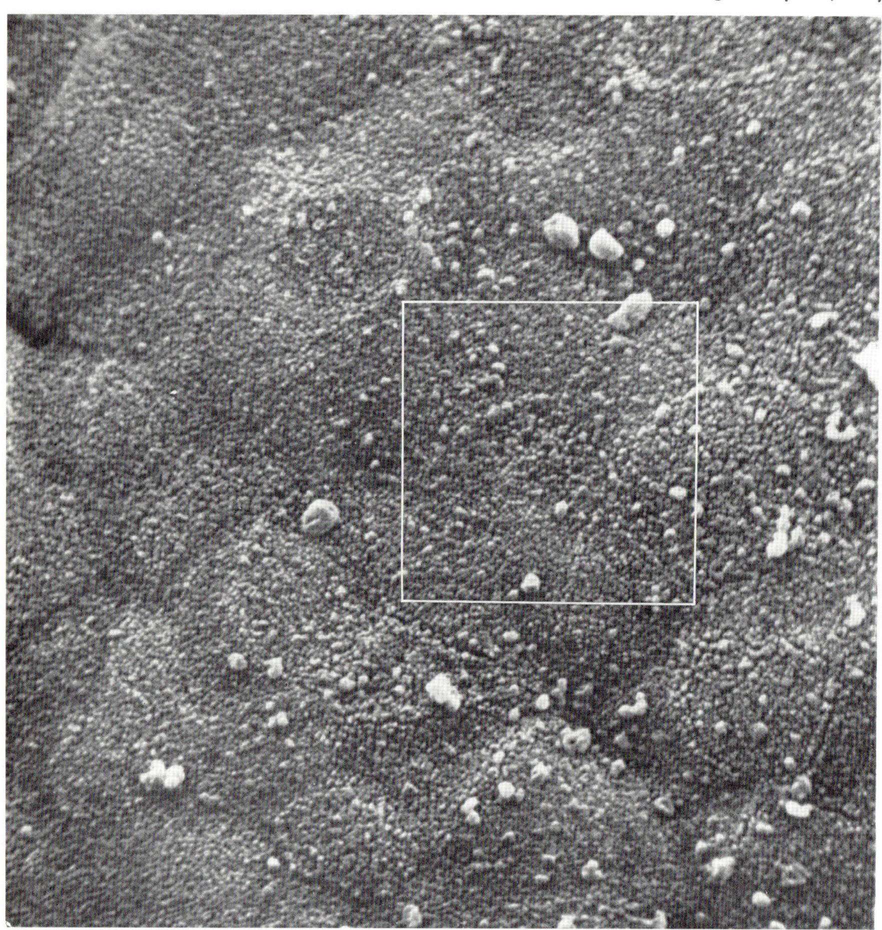

Fig. 22 (×12,000)

Fig. 20 is the light microscopic view of a section of the lymph follicle.

In **Fig. 19** one may identify facets on the dome corresponding to individual epithelial cells which are shown in a closer view in **Fig. 21** (left box in Fig. 19). About thirty cells may be seen here bounded by vague, shallow furrows. The surface of these cells is granular because of fine cytoplasmic protrusions called microvilli. Under high magnification (**Fig. 22**), the short microvilli of these cells appear like scattered pebbles.

Fig. 23 shows the ordinary mucous surface near the lymph follicle and corresponds to the right box in Fig. 19. About fifty cells are seen with their lightly swollen surface. Several flower-shaped structures represent the mucous secretion from individual goblet cells. Other cells of absorptive nature are rough in surface because of thickly-growing microvilli. Under high magnification (**Fig. 24**), approximately five to twenty tiny, round-headed microvilli are stuck together in numerous cauliflower-shaped clumps. It is difficult to obtain a clear figure of individual microvilli of absorptive epithelium under the scanning electron microscope, as they are covered by a polysaccharide-rich "surface coat" which cannot be removed without damaging the fine cellular structure.

Note

The appendix of an adult rabbit was fixed in 10 per cent neutral formalin, dehydrated in acetone, dried in air and simply coated with gold. EM: JSM-2

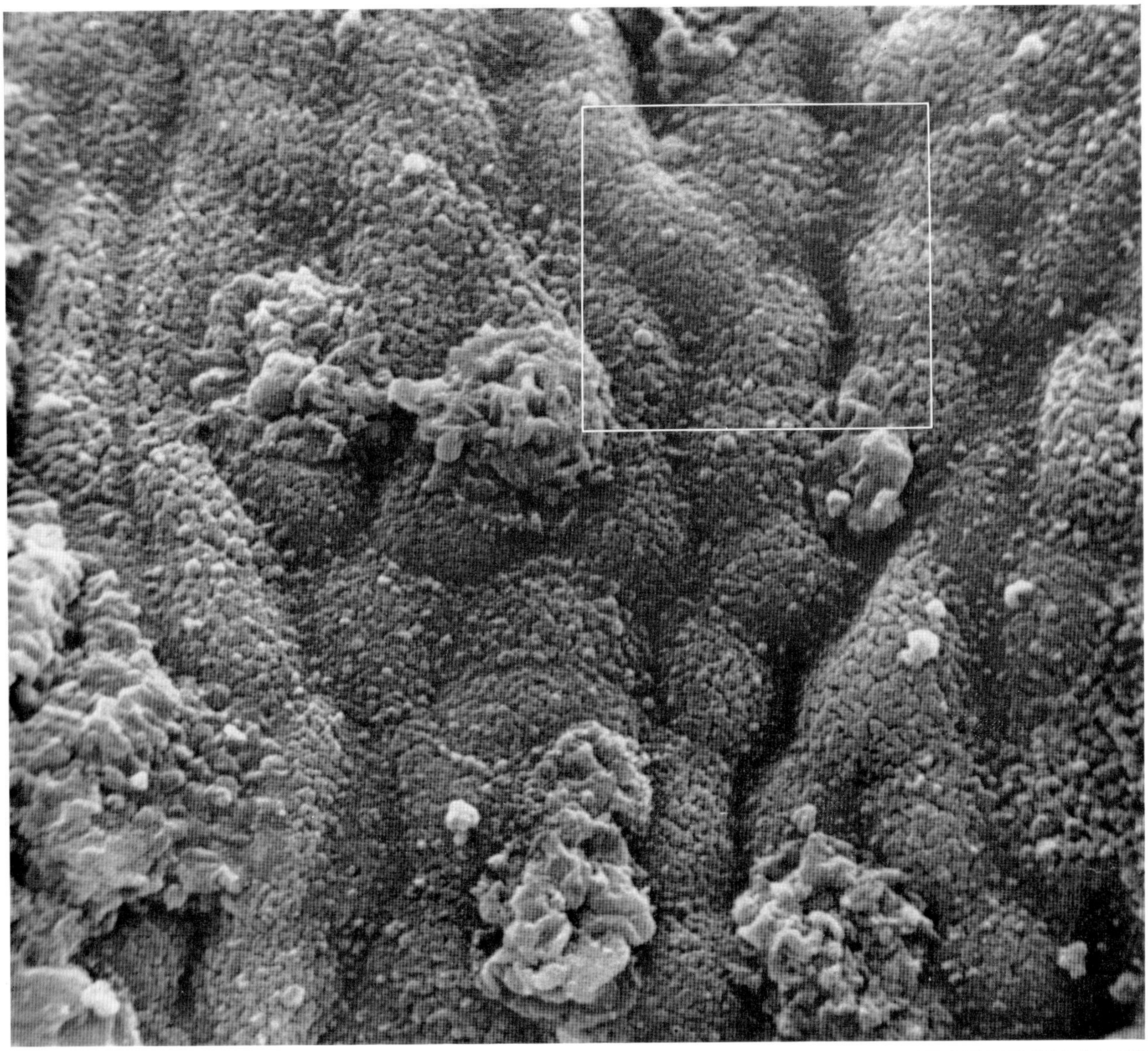

Fig. 23 (×5,100)

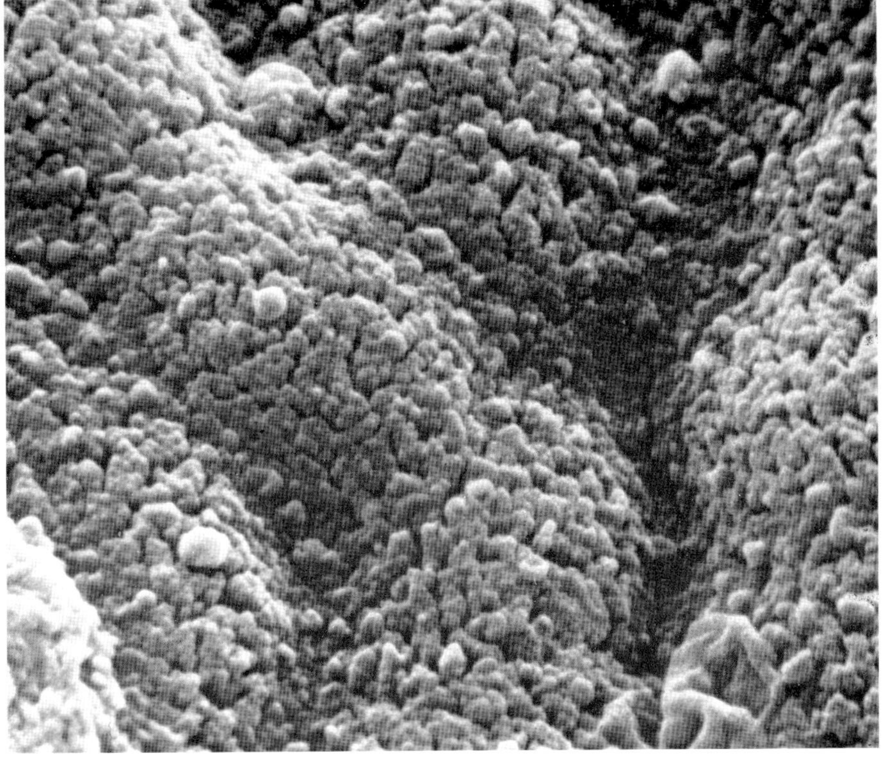

Fig. 24 (×12,000)

6. Hepatic Lobule and Liver Cells

The structural and functional unit of the liver is the hepatic lobule which consists of liver cells and specialized blood capillaries called sinusoids. The liver cells were long believed to form cylindrical cords of double rows of cells, but since ELIAS (1949) investigated them from the view-point of three-dimensional histology, they are known to construct a system of interconnected and fenestrated plates. The sinusoids run through the interspaces of these plates to pour into a central vein, a common drainage penetrating the center of the lobule.

Fig. 25 is a low-power scanning electron micrograph of the rabbit liver. In the right half of this picture the liver has been cut by a razor blade and in the left half it has been broken. In the center of the picture, a central vein is cut obliquely and the profiles of liver cell plates radiate from it. The spaces between the plates which, though not discernible in this picture, contain sinusoids are confluent to the central vein. The

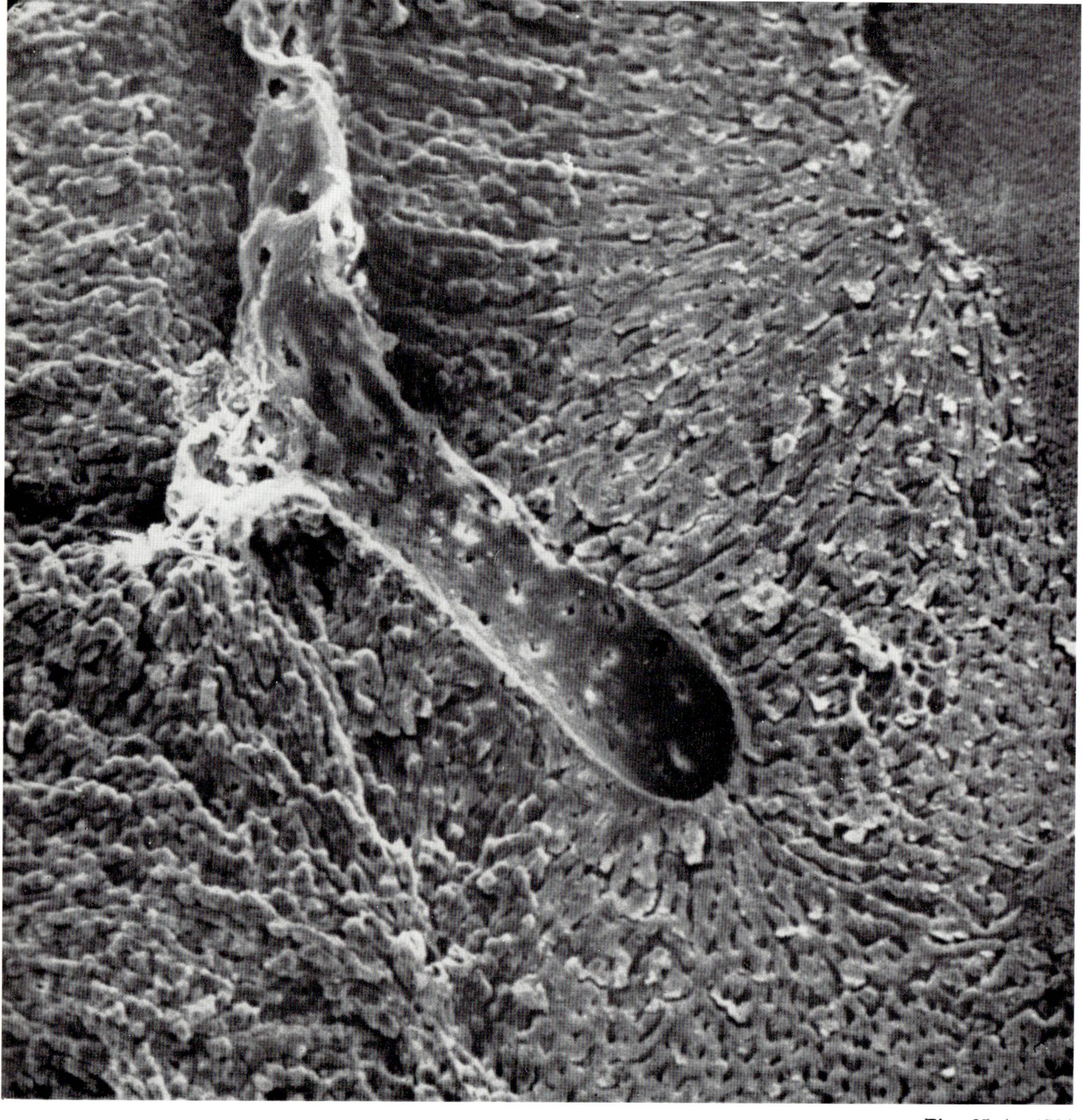

Fig. 25 (× 170)

numerous pores on its wall are the orifices of the sinusoids.

Fig. 26 shows a broken surface of the central part of a lobule. A central vein and the radiation of liver cell plates are seen more closely. The cuboidal and polyhedral units correspond to individual liver cells.

Fig. 27 is a high-power view of liver cell groups in a broken surface. Liver cells were separated at the intercellular boundaries, so that the surfaces which had been hidden by adjacent cells in a natural state were exposed. These surfaces are marked by suture-like double lines and tiny circles as indicated by arrows. They represent bile canalicules which carry the secretion of the liver cells. This micrograph of liver cell groups reminds us of a pile of polyhedral concrete blocks and seems to be compatible with the three-dimensional diagram of liver cells proposed by ELIAS (1949). The gutter-like parts indicated by asterisks are the inner surface of sinusoids thinly covered by coagulated blood plasma.

Note

The liver of the rabbit was perfused with Ringer solution and then with 2.5 per cent glutaraldehyde (0.1M phosphate) through the portal vein. The excised organ, after being kept in the same fixative for a few days, was either cut with a razor blade or broken by hand. Small blocks having an adequate surface were dehydrated in acetone and coated with carbon and gold. EM: JSM-2

Reference

ELIAS, H.: A re-examination of the structure of the mammalian liver. I. Parenchymal architecture. II. The hepatic lobule and its relation to vascular and biliary system. Amer. J. Anat. 84: 311-334; 85: 379-456 (1949).

Fig. 26 (×450)

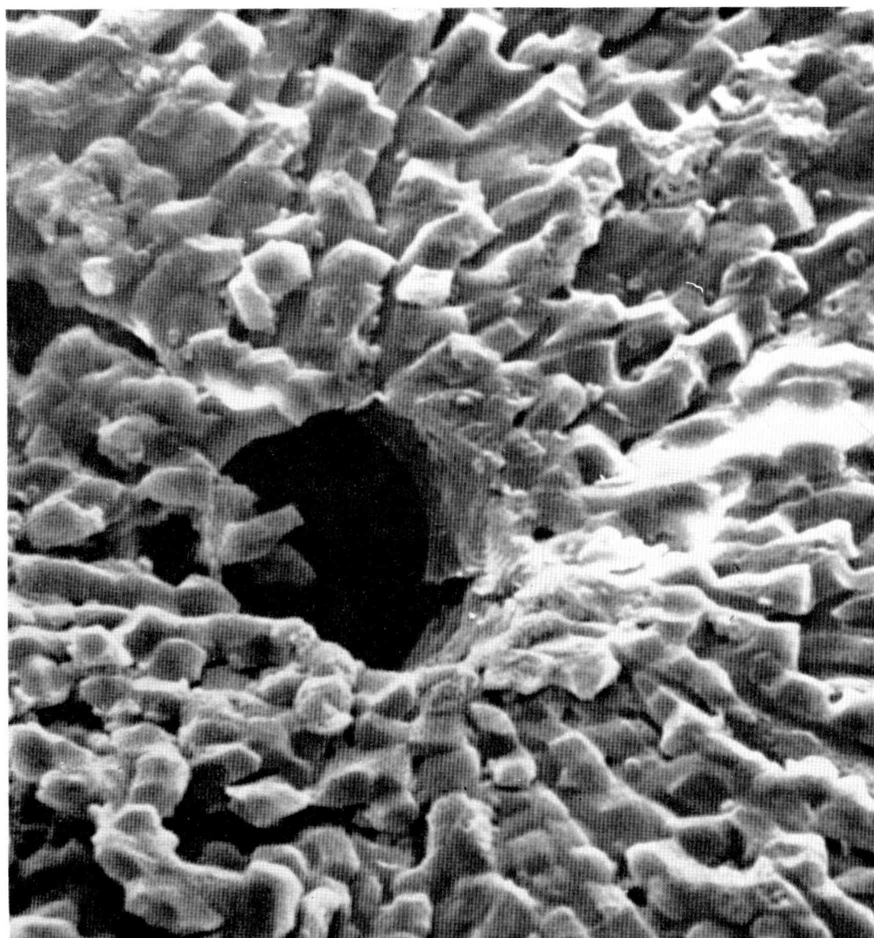

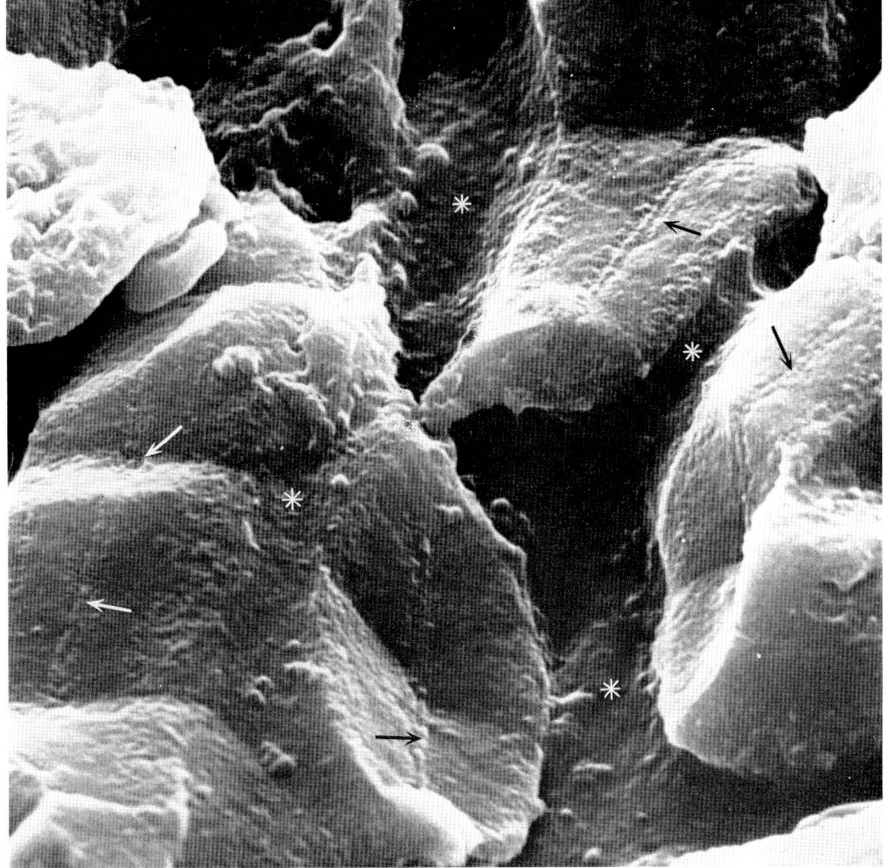

Fig. 27 (×3,600)

7. Renal Corpuscles and Podocytes

The renal corpuscle consists of a glomerulus 0.2–0.3 mm in diameter and a capsule of Bowman loosely encasing it. The glomerulus is a tuft of blood capillaries and is an apparatus for blood filtration. Up to 99 per cent of its filtrate is reabsorbed into the blood while it is transported through the urinary tubule which is connected to the capsule of Bowman, and the remainder is excreted as urine.

In **Fig. 28** showing a cut surface of the rabbit kidney at a low magnification, the glomeruli are each contained in a bowl which corresponds to an opened Bowman capsule. Some bowls appear empty as their glomerulus has dropped during the specimen preparation. The space among the renal corpuscles is filled with sections of urinary tubules.

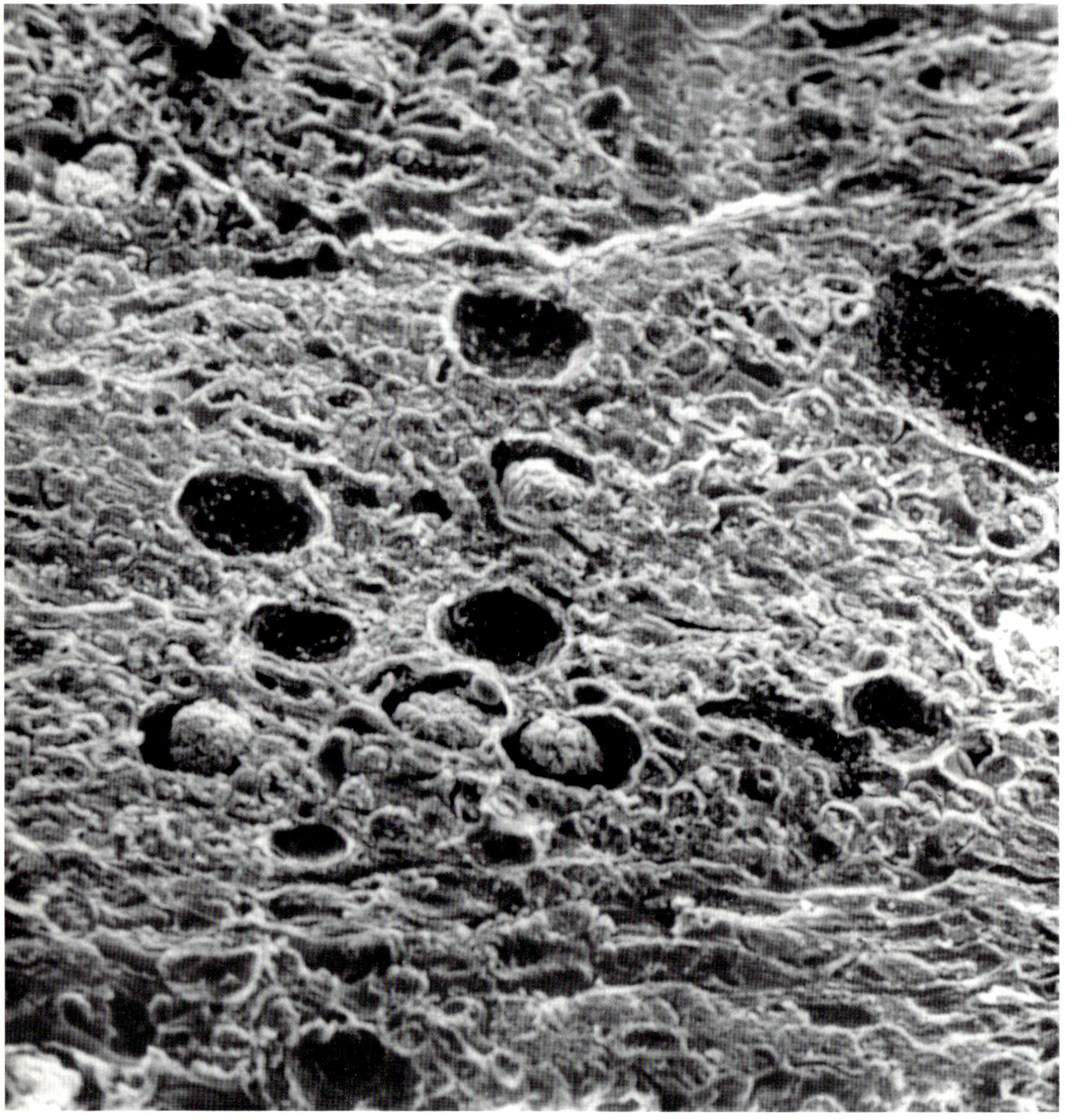

Fig. 28 (× 170)

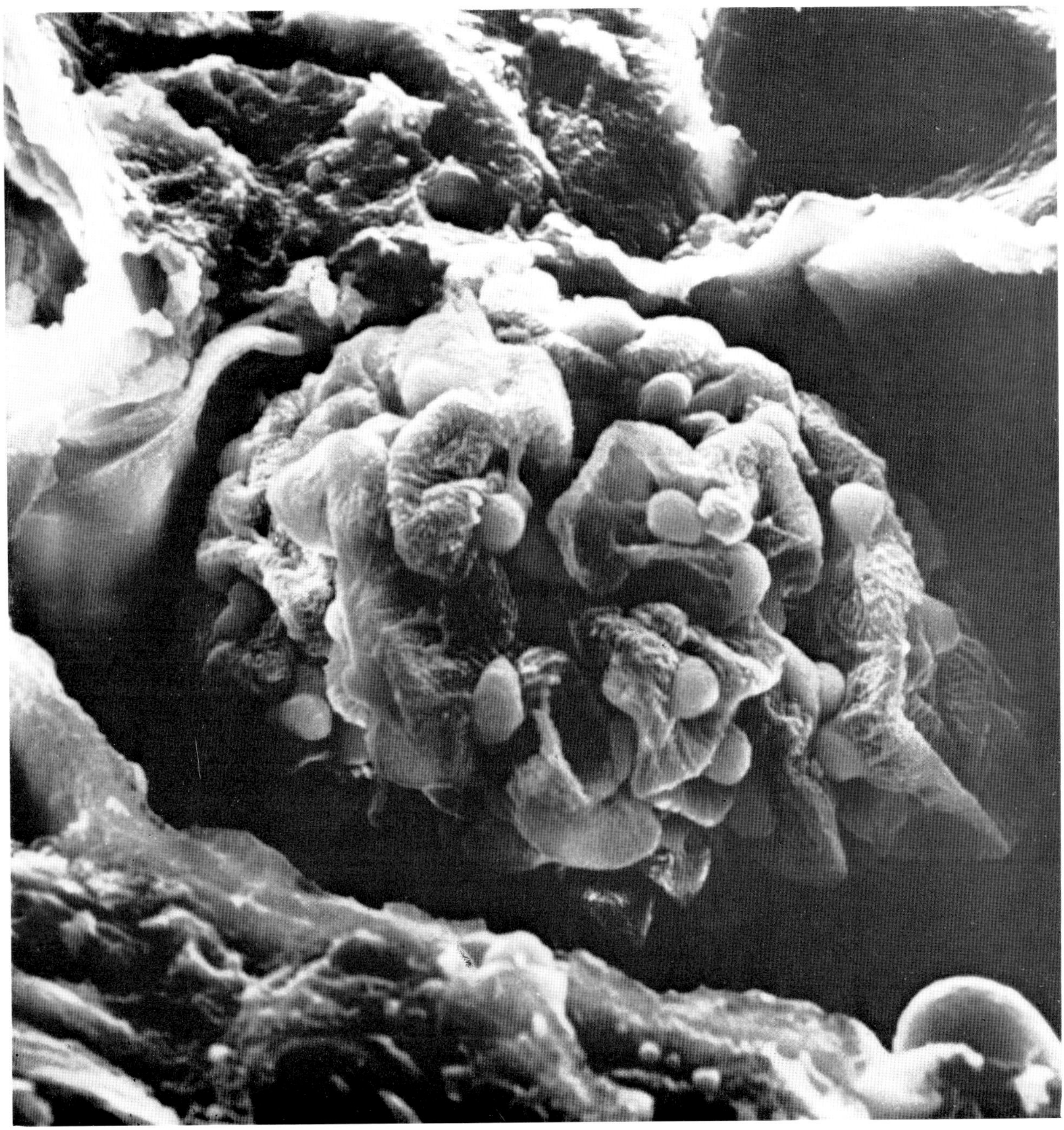

Fig. 29 (×1,700)

One of the corpuscles in Fig. 28 is enlarged in **Fig. 29**. The loops of glomerular capillaries are observed. The round bodies attached to them are the nuclear portion of podocytes, specialized cells of the epithelial lining of the space of Bowman.

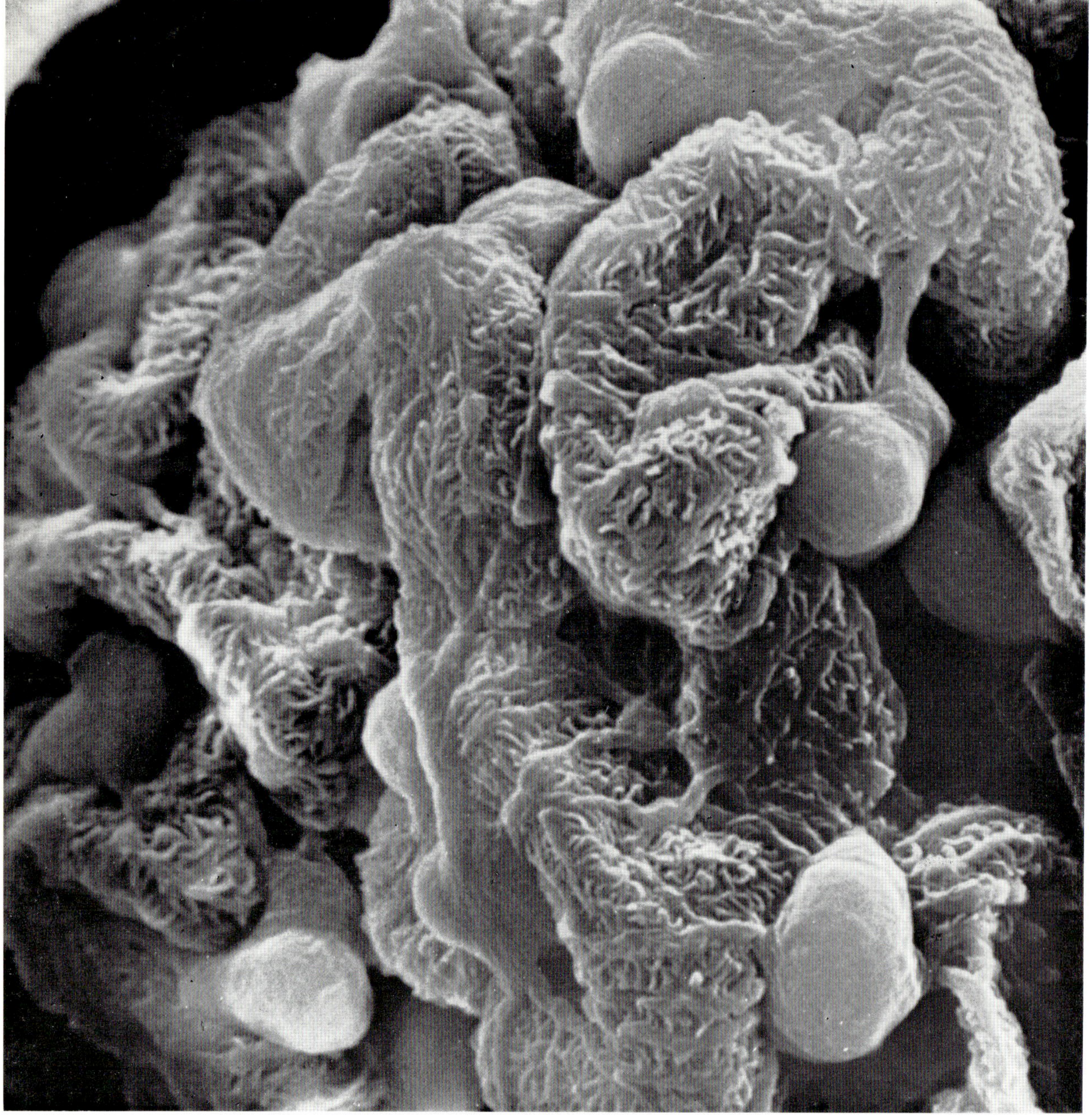

Fig. 30 (×5,100)

A closer view of a part of Fig. 29 shows the form of the podocytes more clearly (**Fig. 30**). These cells may be compared with octopuses whose round heads (nuclear portion) are situated between the limbs of a capillary loop and extend several thick feet (cytoplasmic processes) onto them. As seen under high magnification (**Fig. 31**), these primary prceesses send out, in their turn, more delicate secondary and tertiary processes which wind around the capillary wall. The complex pattern of the ramification of processes reminds us of fern leaves.

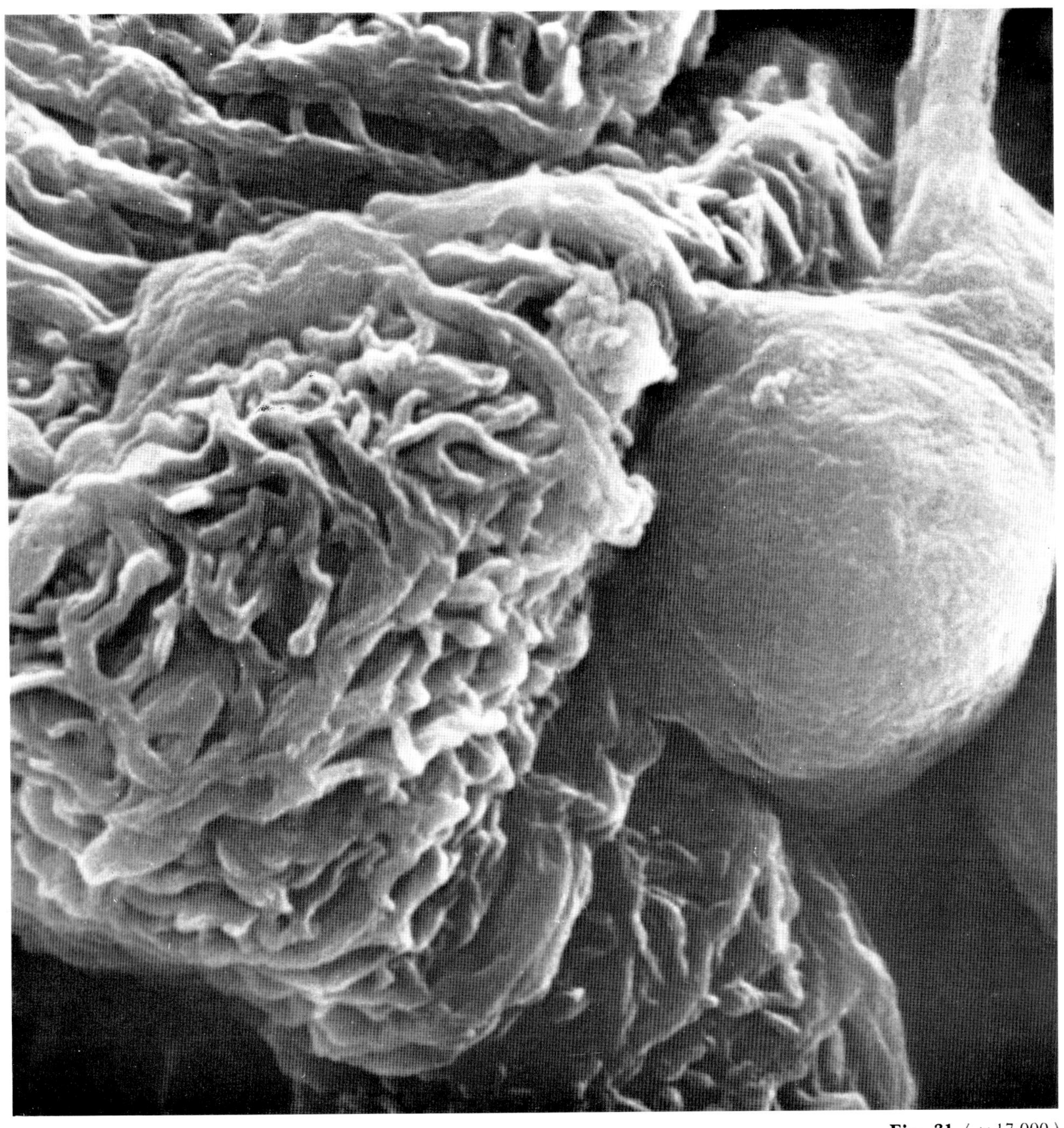

Fig. 31 (× 17,000)

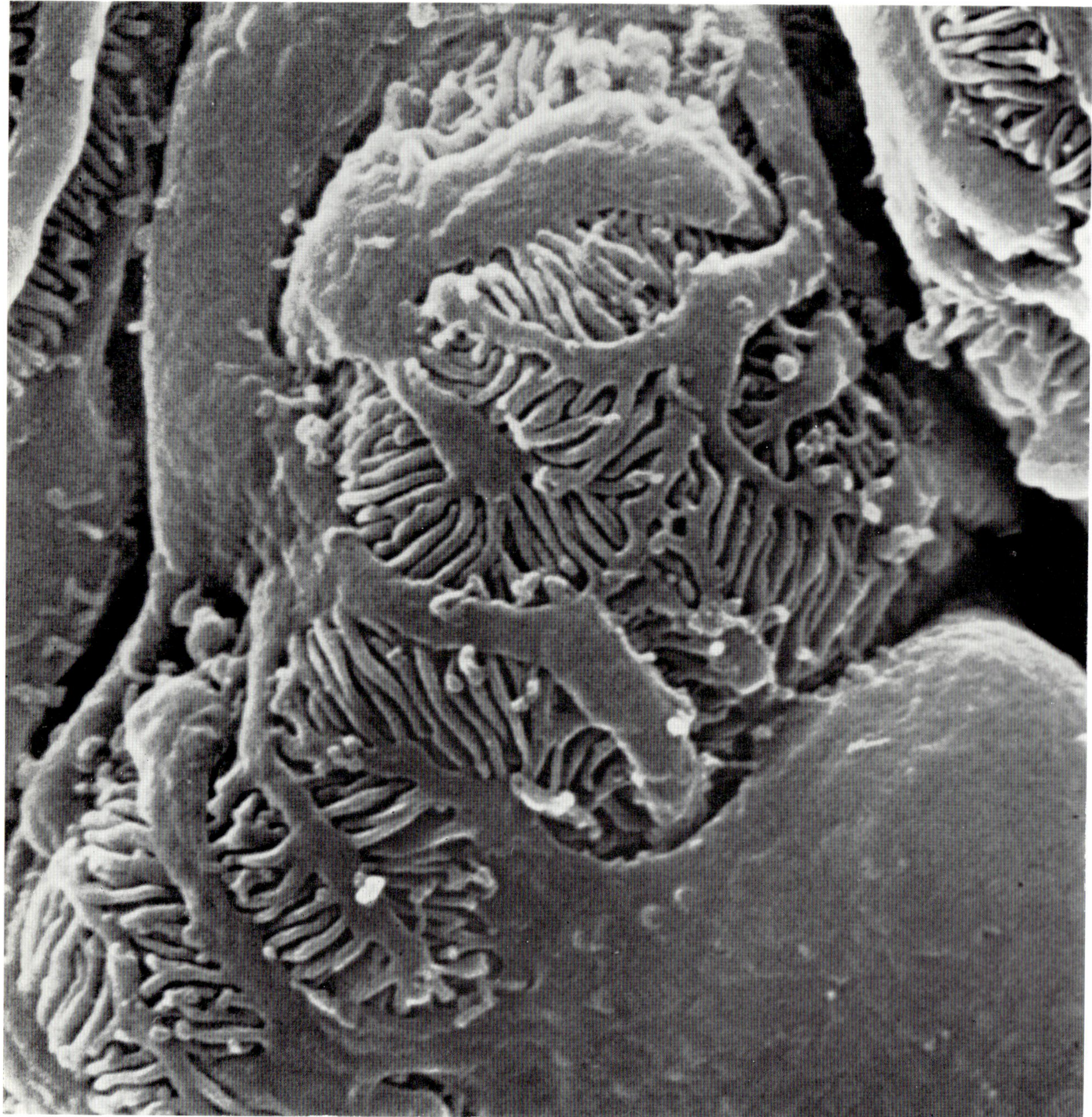

Fig. 32 (×17,000)

The relations of the podocyte processes could be observed more clearly in the glomerulus of the rat. In **Fig. 32** a podocyte is seen at the bottom right corner, while thick processes from other cells come down from the top of the picture. The "fern leaves" from different cell sources are interdigitated with each other. **Fig. 33** shows a part of Fig. 32 in a closer view. It is noticed that small cytoplasmic projections (microvilli) occur on the upper surface of the cell body and processes.

The three-dimensional aspects of the podocytes have been illustrated on the basis of transmission electron micrographs of sections (PEASE, 1955; YAMADA, 1955). The scanning electron microscope, however, reveals the real forms of the cells which differ considerably from the figures hitherto proposed, especially in the ramification pattern of the feet and their mutual topological relationship. Application of this electron

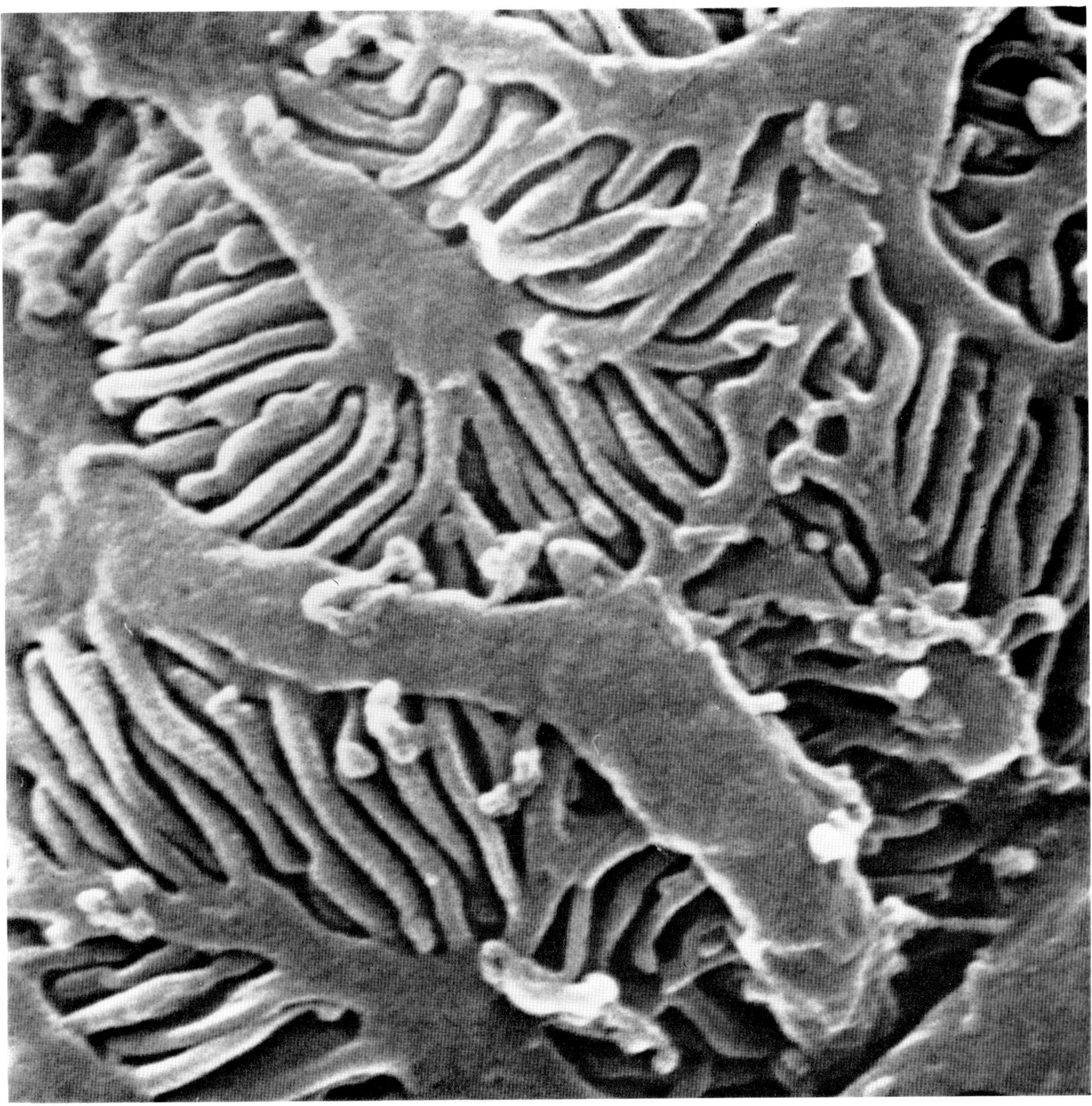

Fig. 33 (×34,000)

microscope to the analysis of the developmental, as well as pathological, changes of the podocytes seems to promise a fruitful field of study.

Fig. 34 shows a capsule of Bowman of the rabbit with its sink leading to the urinary tubule. The glomerulus in this picture is sectioned and the capillaries opened. In an opened capillary as shown in **Fig. 35** one may observe the fine structure of their inner surface. **Fig. 36** shows at a high magnification the inner surface of capillary endothelium of the rat. The thin cytoplasm of the endothelial cell runs as narrow ridges and the attenuated web-like portions between them are provided with numerous tiny pores. Between this pored endothelium and the feet of the podocytes is sandwiched a basement membrane, a kind of filter layer, which cannot be observed by scanning electron microscopy (**Fig. 37**).

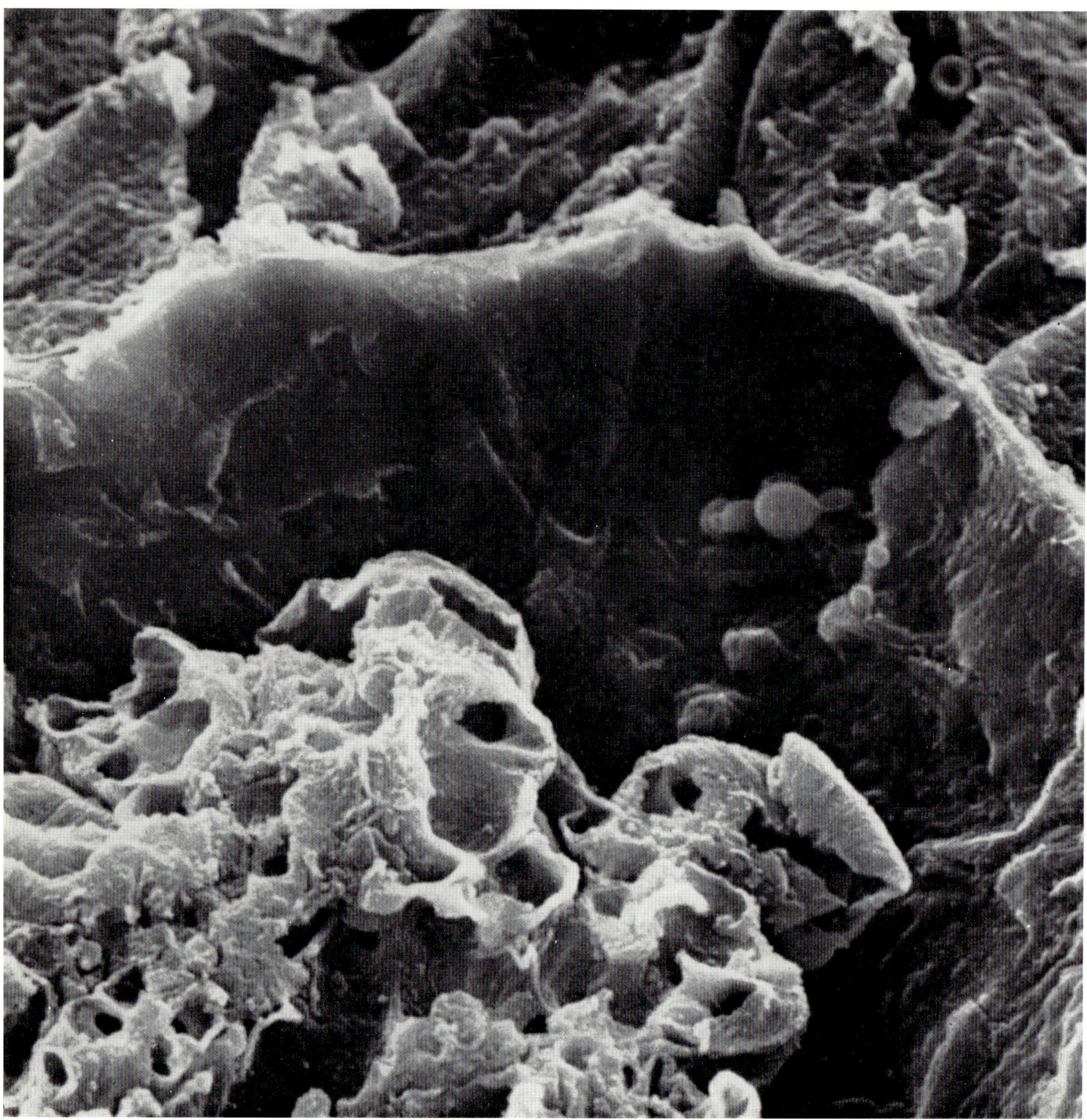

Fig. 34 (× 1,700)

Note

The kidney of the rabbit was fixed by an infusion via renal artery of 2.5 per cent glutaraldehyde (0.1M phosphate buffer). A few days later, the kidney was cut with a razor blade into small blocks each having an appropriate cut surface. They were dehydrated in acetone, dried in air and coated with carbon and gold. EM: JSM-2

Reference

ARAKAWA, M.: Scanning electron microscopy of the rat glomerulus. Lab. Invest. (1970, in press).

BUSS, H. and W. KRÖNERT: Zur Struktur des Nierenglomerulum der Ratte. Rasterelektronenmikroskopische Untersuchungen. Virchows Arch. Abt. B, Zellpathol. 4: 79-92 (1969).

FUJITA, T., J. TOKUNAGA and M. MIYOSHI: Scanning electron microscopy of the podocytes of renal glomerulus. Arch. histol. jap. 32: 99-113 (1970).

MORGENROTH, K.: Comparative examinations of concrements of the human kidney by scanning electron-microscopy. Virchows Arch. Abt. A, Pathol. Anat. 347: 33-43 (1969).

PEASE, D. C.: Fine structures of the kidney seen by electron microscopy. J. Histochem. Cytochem. 3:295-308 (1955).

YAMADA, E.: The fine structure of the renal glomerulus of the mouse. J. biophys. biochem. Cytol. 1: 551-566 (1955).

Figs. 28, 32, 33 and 36 by courtesy of the Arch. histol. jap.

Fig. 35 (× 12,000)

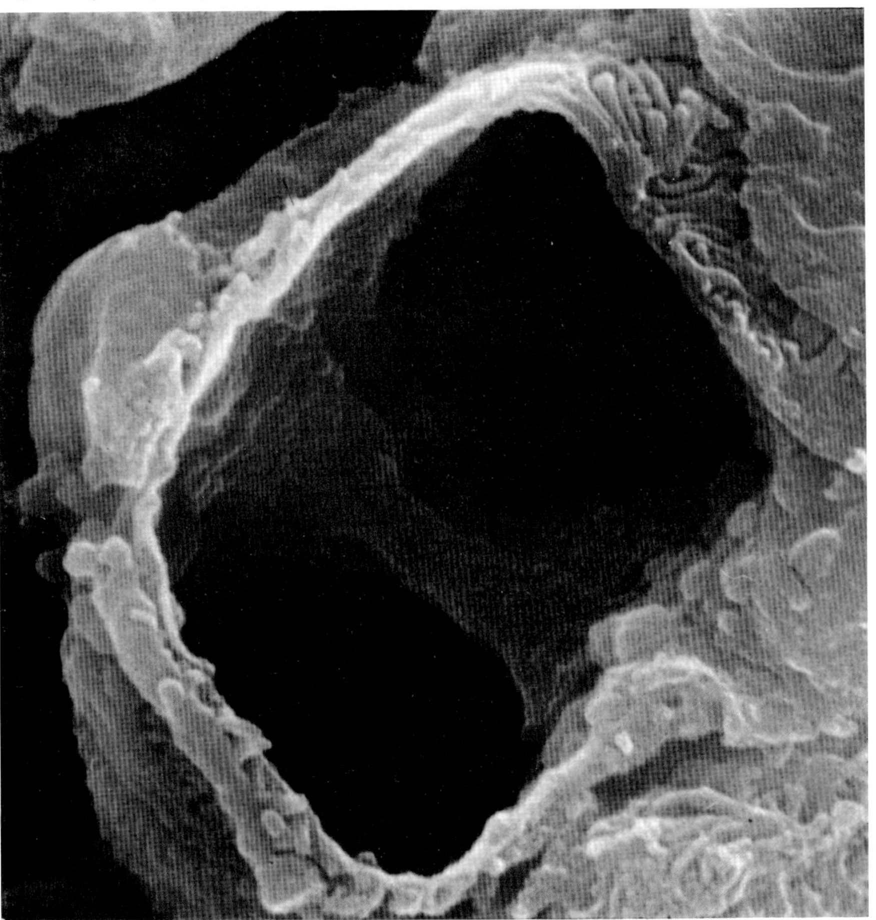

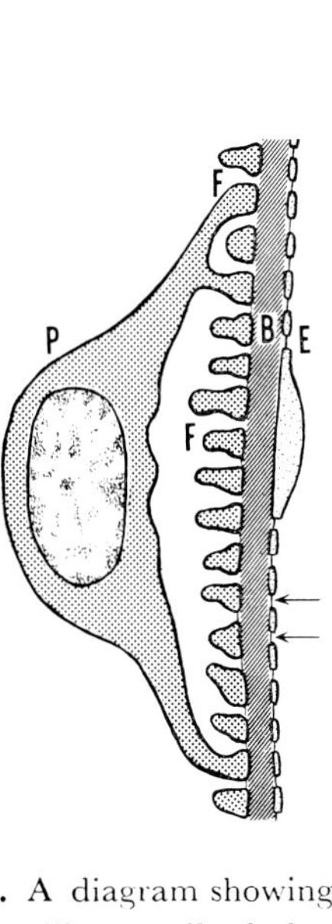

Fig. 37. A diagram showing the section of the capillary wall of glomerulus.

B: basement membrane. **P:** podocyte with its foot-processes **(F)**. **E:** attenuated endothelial cytoplasm with pores (arrows).

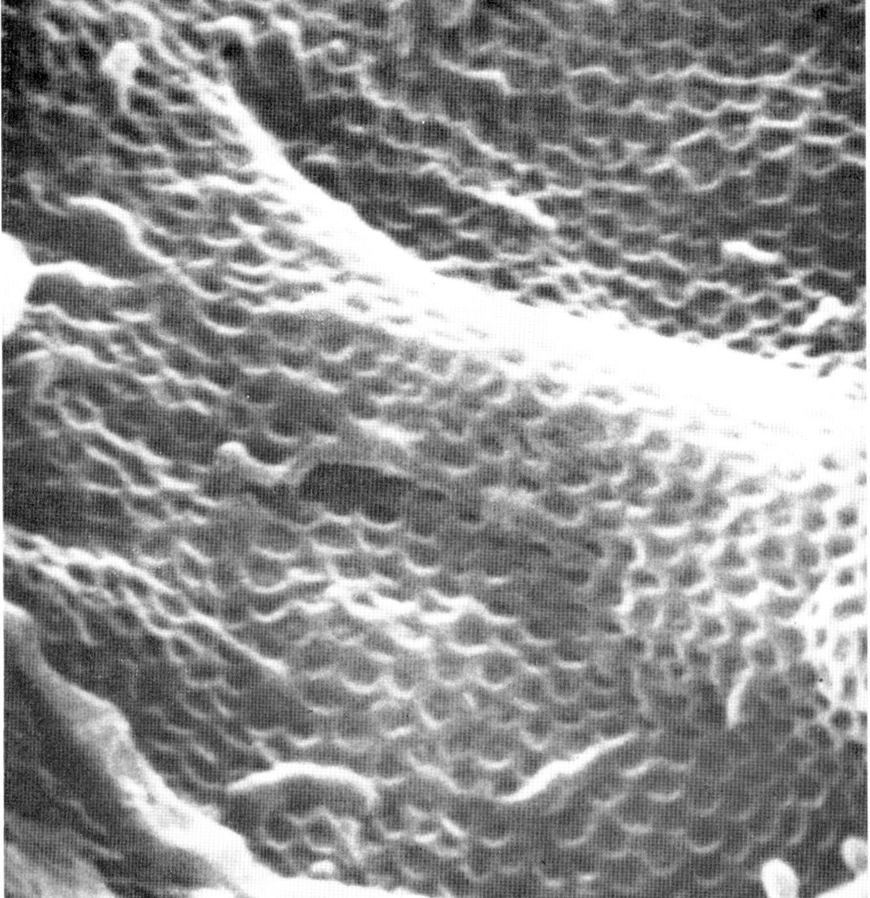

Fig. 36 (× 45,000)

8. Cilia in the Trachea

In the human body ciliated epithelia occur in the nasal cavity, trachea and bronchi as well as in the oviduct. The apical surface of the epithelial cells is covered by cilia which are motile processes of cytoplasm and have a rapid to and fro beat. In the case of the respiratory tract, the effective stroke of the ciliar movement is directed upwards, so that dust and mucus are consistently transported toward the mouth and nose.

Fig. 38 is a low-power scanning electron micrograph of the inner surface of the rabbit trachea showing innumerable cilia waving as in a windy grain field. **Figs. 40** and **41** are closer views of those cilia. Though their surface may appear transversely or obliquely banded after single fixation in formalin or glutaraldehyde (**Fig. 39**) as pointed

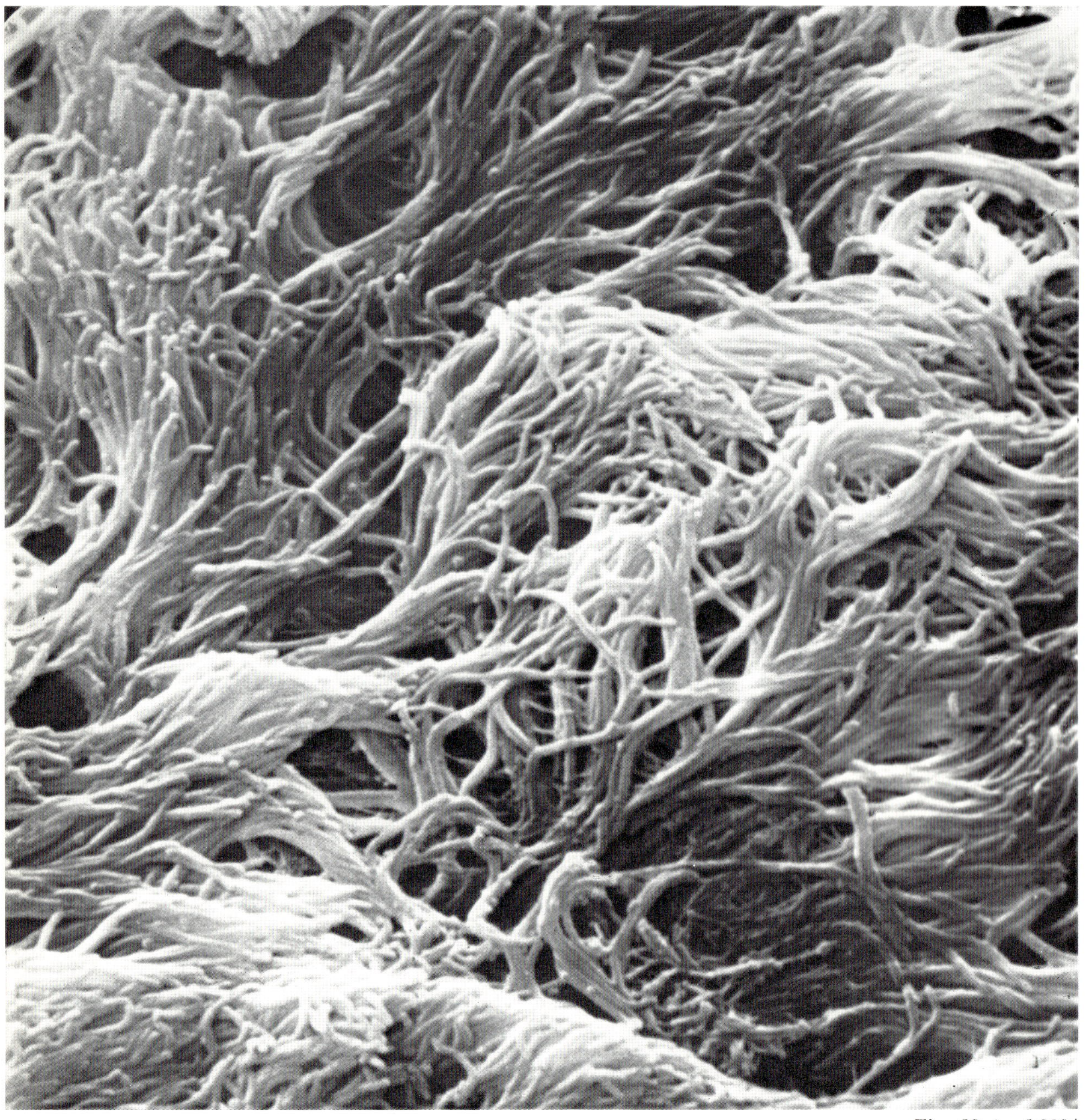

Fig. 38 (×6,000)

Fig. 39 (× 30,000)

Fig. 40 (× 24,000)

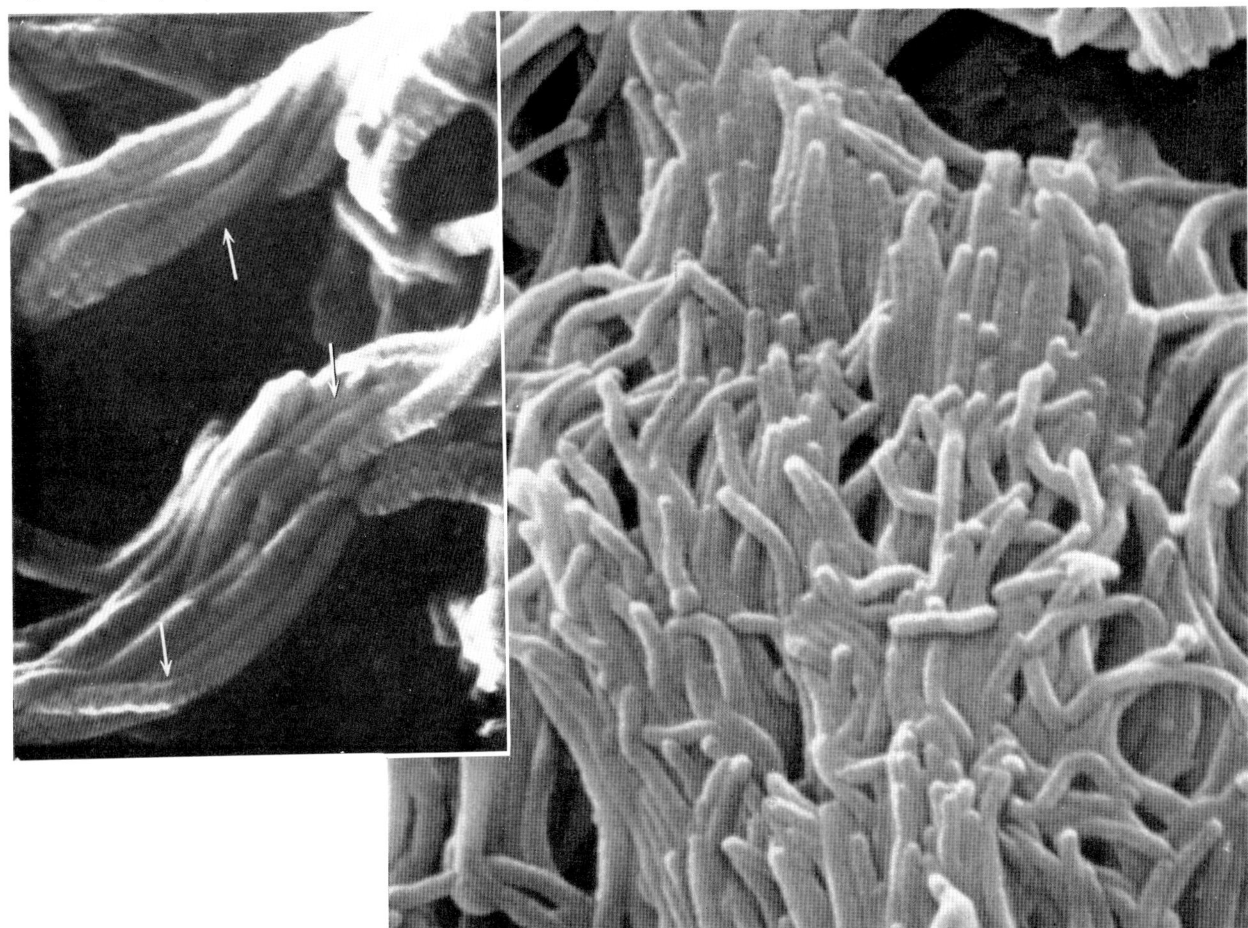

out by Sakaguchi and Watanabe (1969), this seems to be artifact as the ciliar surface appears smooth when the aldehyde-fixed specimens are hardened by postfixation in osmium tetroxide (Figs. 40 and 41). The end of the cilia is never tapered but is blunt like the cut end of noodles. The small balls seen in Fig. 41 seem to be torn parts of the cilia.

Note

The inner surface of excised pieces of the rabbit trachea was carefully and patiently washed with physiological saline. The pieces were fixed in 2.5 per cent glutaraldehyde (0.1M phosphate buffer) and then in 2 per cent osmium tetroxide, dehydrated in acetone, dried in air and coated with carbon and gold. EM: JSM-2

Reference

Barber, V. C. and A. Boyde: Scanning electron microscopic studies of cilia. Z. Zellforsch. 84: 269-284 (1968).

Sakaguchi, H. and S. Watanabe: Specimen preparation for scanning electron microscopy (in Japanese). Saishin Igaku 24: 2544-2550 (1969).

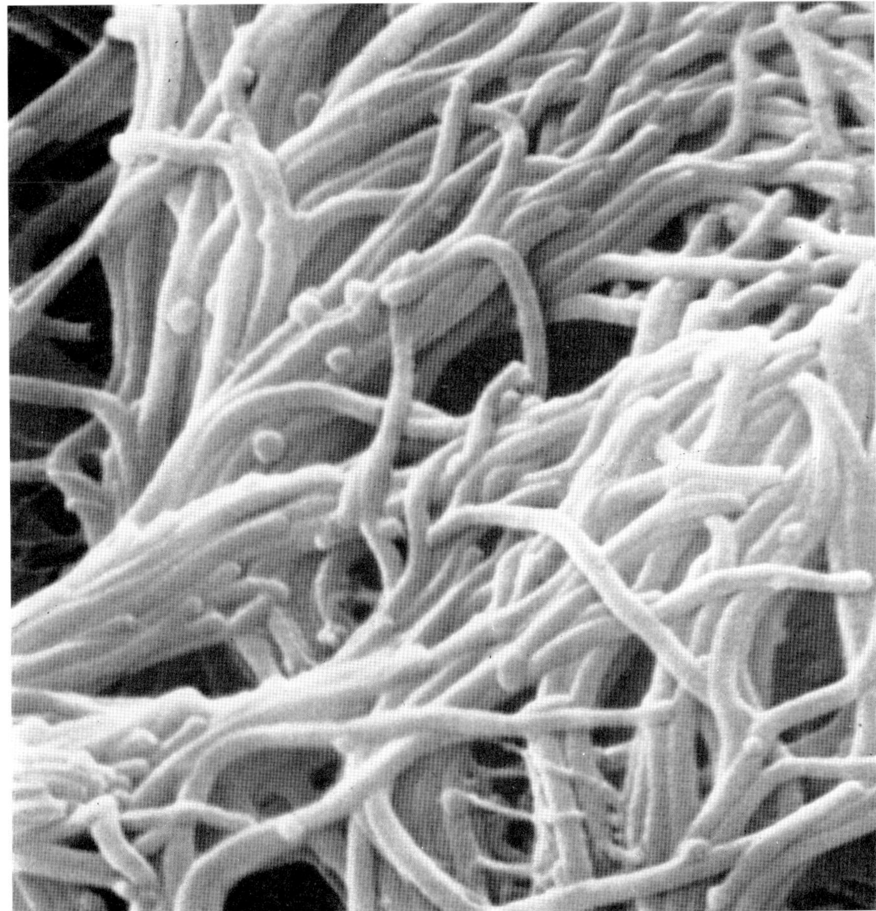

Fig. 41 (× 24,000)

9. Lung and Its Alveoli

In the lung the bronchus branches repeatedly into smaller ducts to lead the air to the tiny end sacs called alveoli. The gas exchange takes place through the delicate wall of the alveoli. These structures which appear only labyrinth-like in microscopic sections are clearly visualized in their three-dimensional continuity when the cut surfaces of the organ are observed under the scanning electron microscope. Observations by this method, though only at very low magnifications, have been published by an American research group (JAQUES et al., 1965).

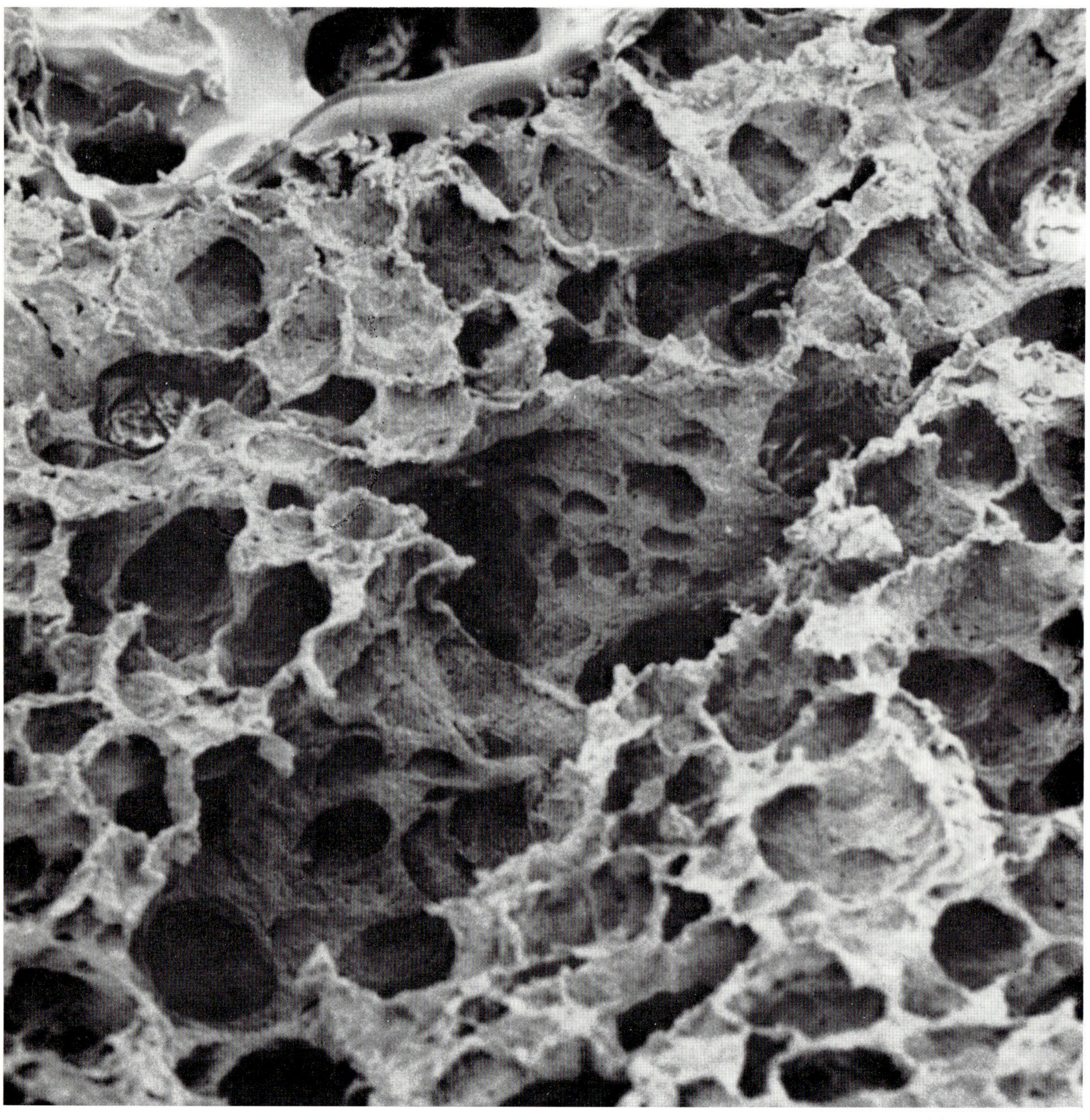

Fig. 42 (×510)

Fig. 43 (× 1,200)

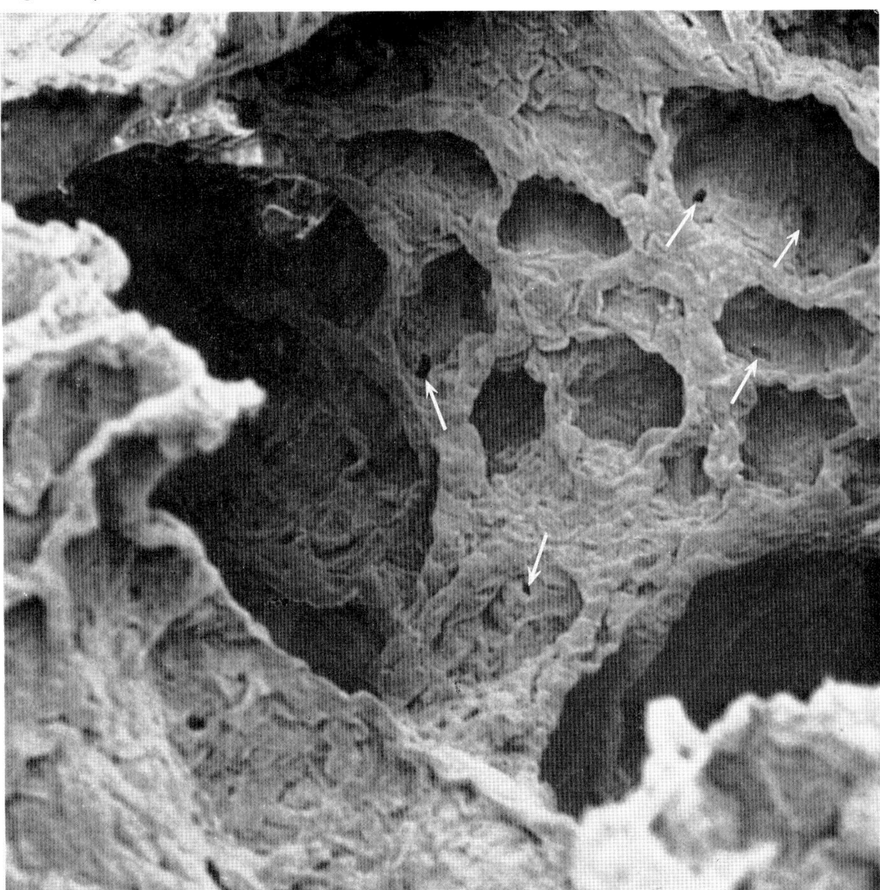

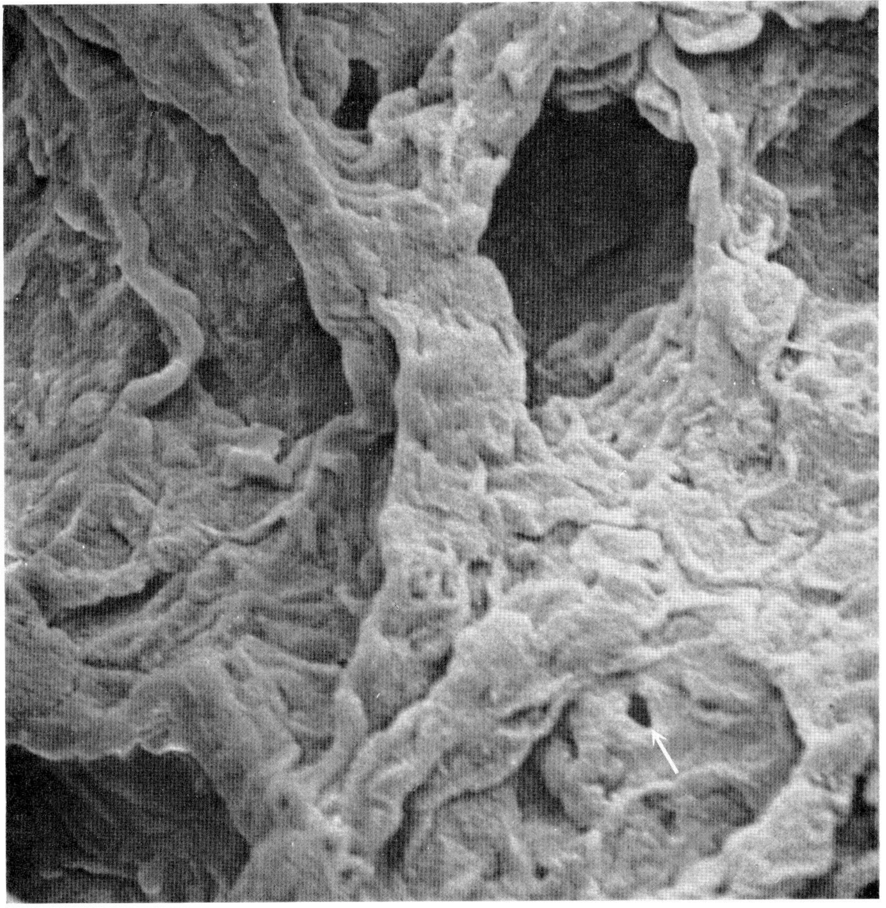

Fig. 44 (× 3,600)

Fig. 42 shows a cut surface of the rabbit lung. In the center of the picture an elongated cavity called alveolar duct is seen. This duct corresponds to the bronchial termination and receives about a hundred alveoli on its sides and at its end. As visualized in this micrograph, the alveolar duct may be compared with a corridor while the alveoli are like doorless rooms bordering it.

Figs. 43 and **44** are closer views of this part. Worthy of note are the small openings in the septa of the alveoli (arrows). The existence of these "alveolar pores" has been much disputed on the basis of the observations of sections. The twisted strings in relief on the alveolar wall likely correspond to the blood capillaries running directly beneath the respiratory epithelium. The epithelial cells, especially those of a large type, may possibly be represented by the ovoid or irregular-shaped elevations on the wall surface.

Note

The lung of a rabbit was excised and injected from the cut end of the bronchus with 10 per cent formalin until the natural volume of the lung was reached. The bronchus then was ligated to prevent the escape of the injected fluid. After being sustained in 10 per cent formalin for several days, the organ was cut with a razor blade into appropriate blocks. They were dehydrated in acetone, dried in air and coated with carbon and gold. EM: JSM-2

Reference

JAQUES, W. E., J. COALSON and A. ZERVINS: Application of the scanning electron microscope to human tissues. Exp. molec. Pathol. 4: 576-580 (1965).

10. Sinuses in the Spleen

The spleen contains two different structures of different functions. One structure is called white pulp and is represented by small nodes or Malpighian corpuscles dispersed in the organ. They are lymphoid tissue and generate lymphocytes which may produce and carry antibodies. The other part of the spleen is the red pulp which is formed of a sponge-like tissue containing ample blood. The red pulp is the site of destruction of aged blood cells. It consists mainly of specialized venous capillaries called splenic sinuses and a network of reticular cells called splenic cords or Billroth cords. The scanning electron micrographs shown here are all concerned with the red pulp with special reference to the fine structure of sinus walls.

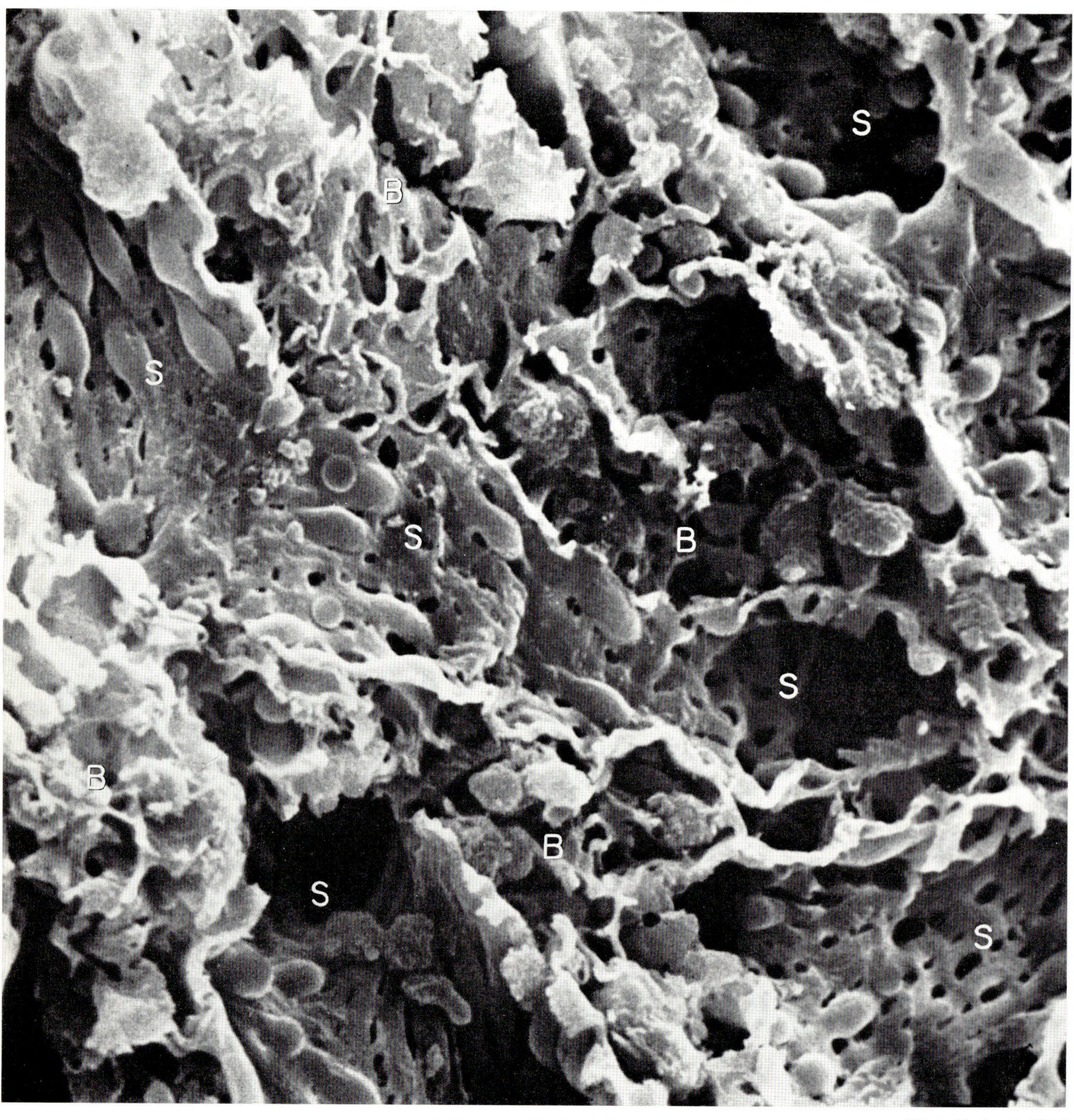

Fig. 45 (×1,700)

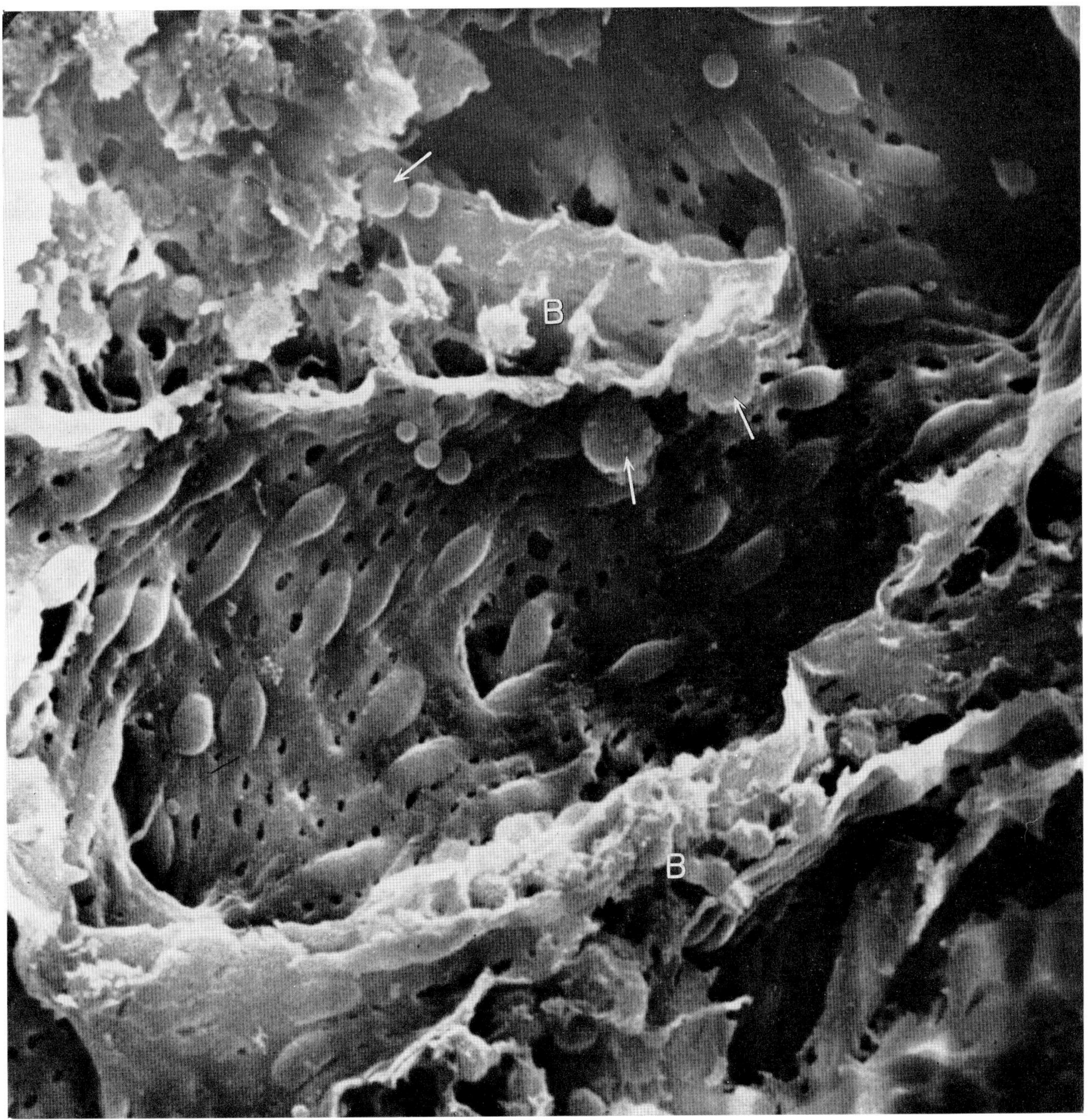

Fig. 46 (× 1,700)

Fig. 45 shows a general view of the red pulp of the rabbit. The corridors and tunnels labeled "S" are sinuses whose walls are lined with a lattice-like cell layer. The sponge-like parts labeled "B" correspond to the cords of Billroth.

Fig. 46 shows the inner surface of a U-shaped sinus. The lattice structure of its wall with pores is obvious. The spindle-shaped swellings on the wall are the nuclear portions of the cells lining the sinus.

Although the red pulp is usually heavily infiltrated with red and white cells of circulating blood, they have been washed away in this specimen except for a few remaining cells of a round shape (arrows). The walls of the sinuses and the meshes of Billroth cords have been thus exposed.

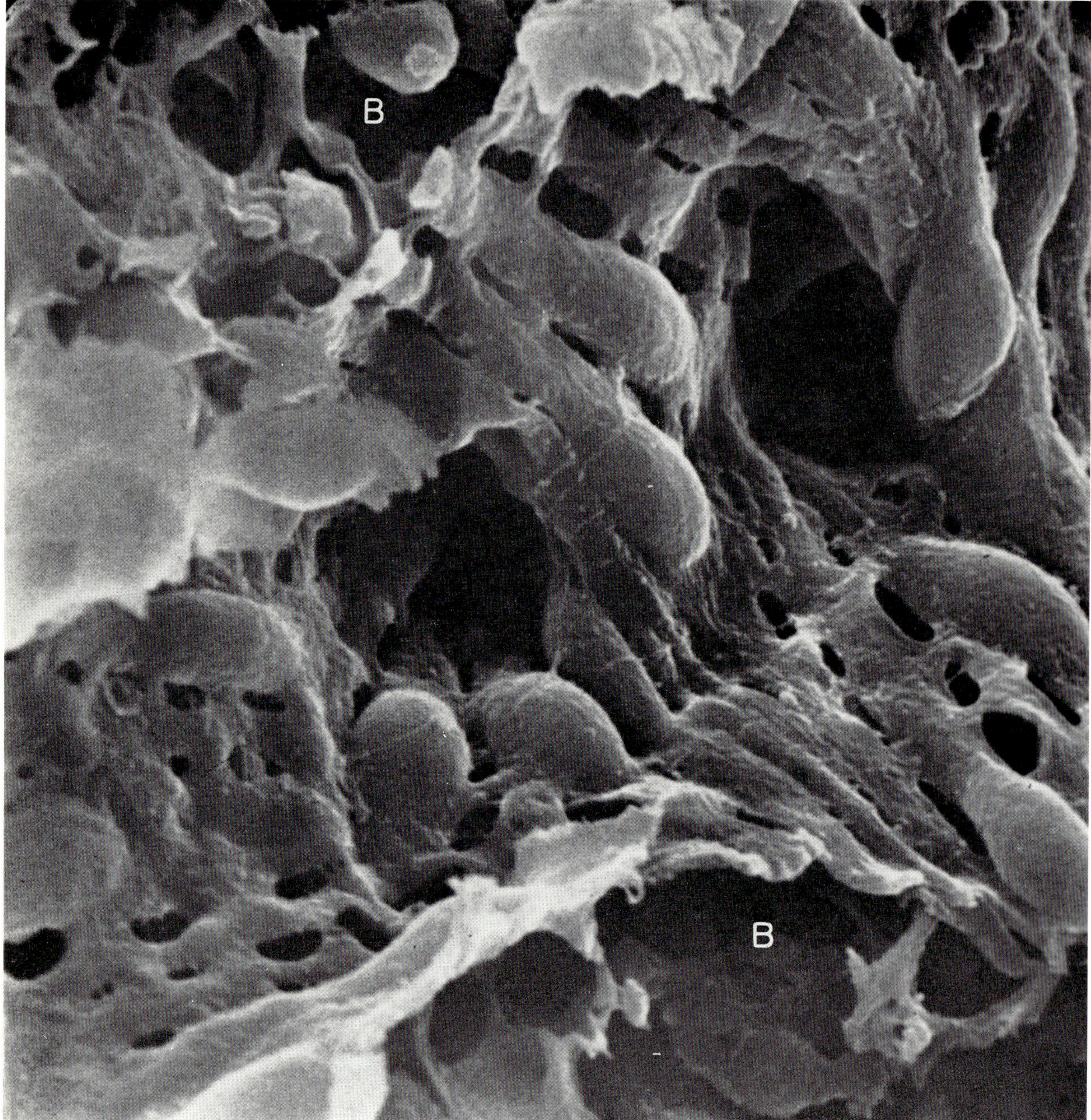

Fig. 47 (×3,600)

Fig. 47 shows at higher magnification a branching portion of a sinus. The lumen of the sinus is continuous to the labyrinths of Billroth through the pores of the lattice. The significance of the red pulp of the spleen as a filter of the blood is visualized in this and other scanning electron micrographs.

Fig. 48 (× 12,000)

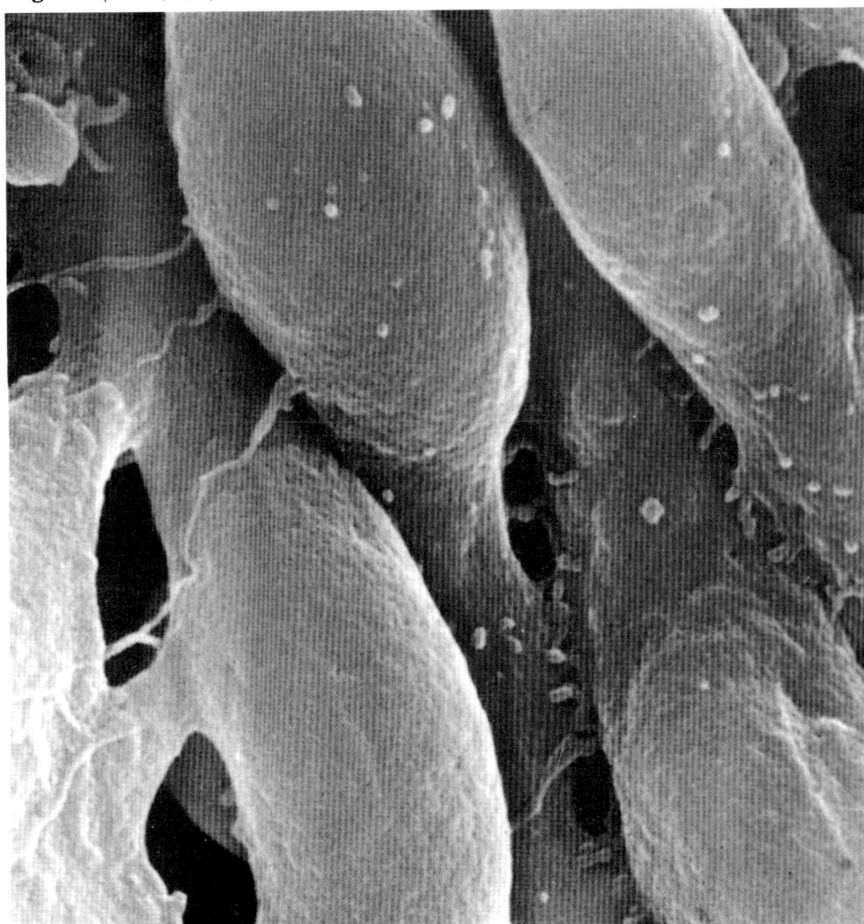

Figs. 48 and **49** are close-up views of the sinus wall. The cells forming a palisade are conspicuously swollen around their nucleus. Belt-like cytoplasm extends from both ends of the nucleus. Neighboring cells are combined by lateral short processes, thus bounding the elongate perforations of an apparently permanent nature in the wall. The occurrence or non-occurrence of permanent openings in the sinus wall and, if they are present, of junctions of lateral processes of the lining cells has been much disputed.

Figs. 48 and 49 further reveal the occurrence of microprojections on the nuclear and peripheral portions of the cell. Some are like tiny warts and others deserve to be called microvilli. There also occur long strings of cytoplasm creeping like ivy along the lattice.

Note

The spleen of the rabbit was thoroughly perfused with Ringer solution and then with 2.5 per cent glutaraldehyde (0.1 M phosphate buffer) via splenic artery. The spleen, excised and cut into pieces, was kept in the same fixative, dehydrated in acetone, dried in air and coated with carbon and gold. EM: JSM-2

Reference

MIYOSHI, M., J. TOKUNAGA and T. FUJITA: The red pulp of rabbit spleen studied by the scanning electron microscope. Arch. histol. jap. 32: 289-306 (1970).

Figs. 45-49 by courtesy of the Arch. histol. jap.

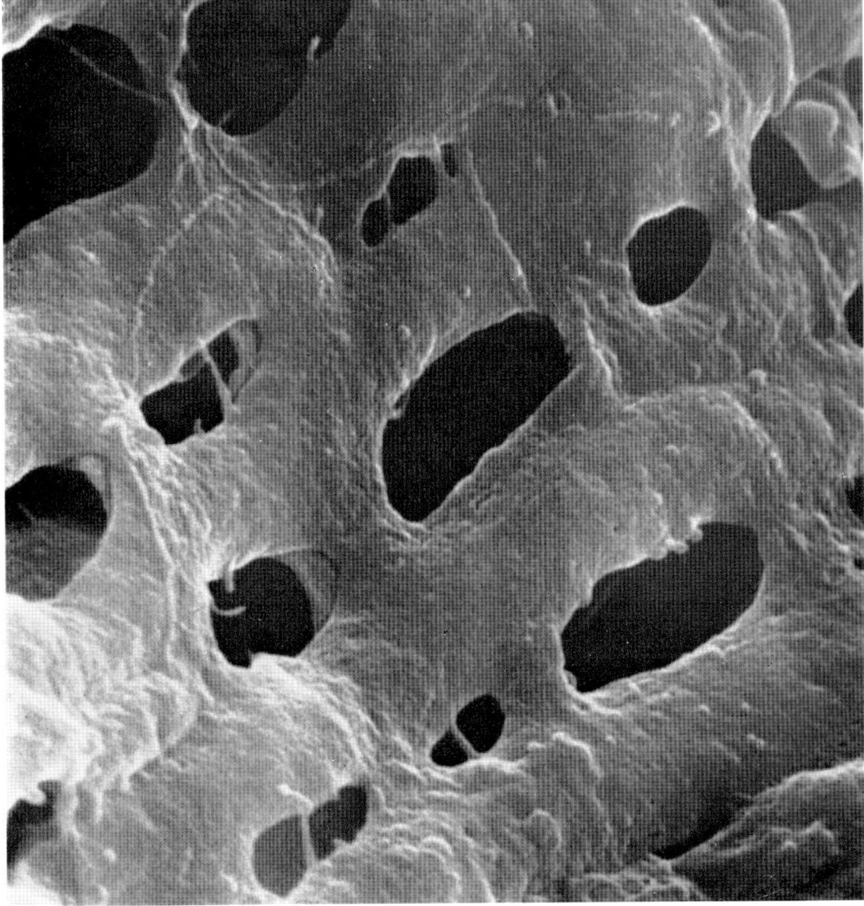

Fig. 49 (× 12,000)

TISSUES OF THE JOINT

11. The Synovial Membrane of the Knee Joint

The synovial membrane is a relatively soft connective tissue lining the joint cavity. Its inner surface is covered by tall cells called superficial cells. They do not form a continuous layer as epithelial cells do, but keep clefts between themselves so that the connective tissue matrix beneath them directly communicates with the joint fluid through these spaces.

Fig. 50 shows the inner surface of the normal human knee joint at a low magnification. The large elevations of the synovial membrane called synovial plicae are covered with innumerable protrusions of variable size which correspond to the superficial cells.

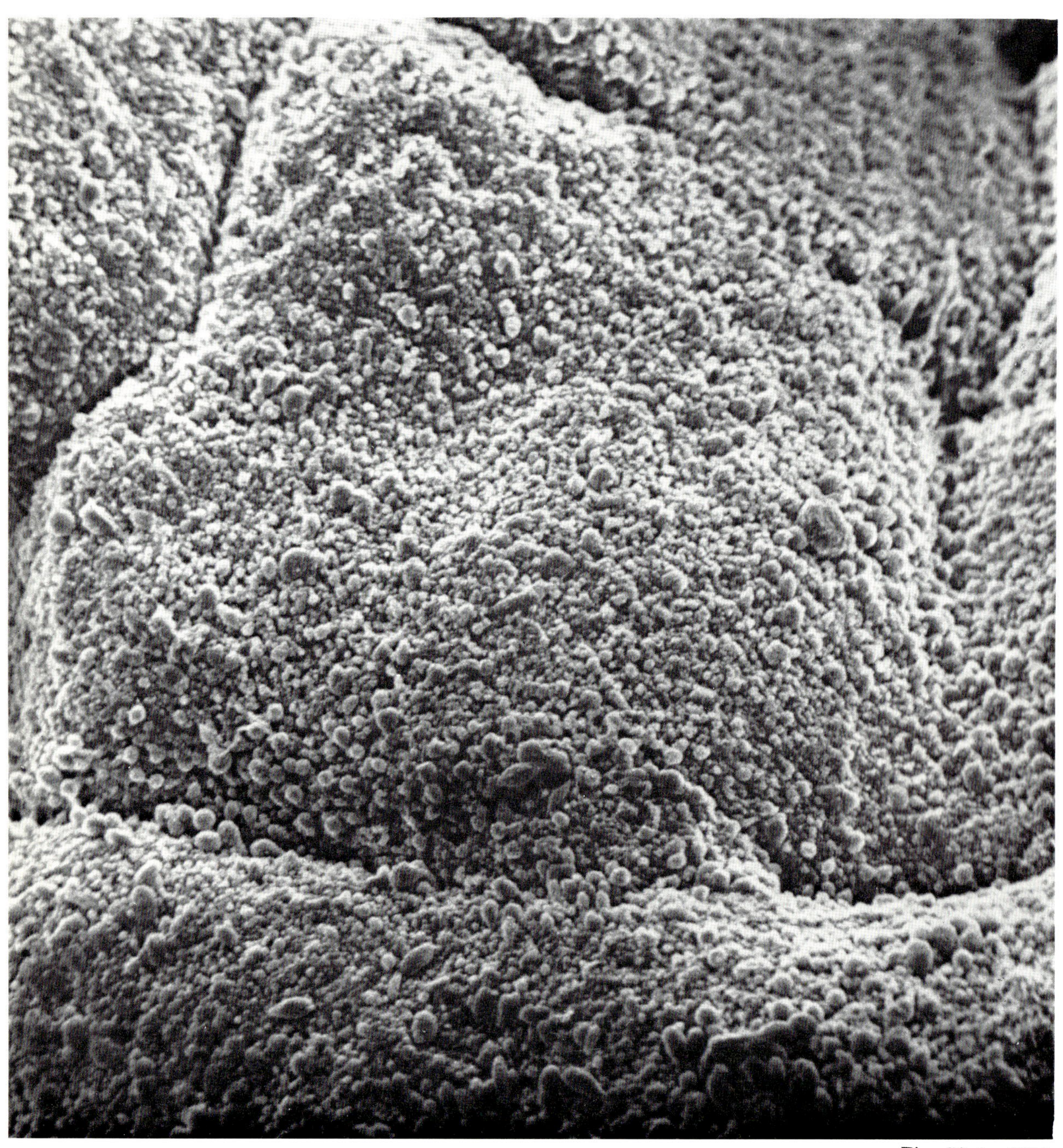

Fig. 50 (×510)

Fig. 51 (×3,600)

Fig. 51 shows these cells in a closer view. Individual cells, seen in a finger-tip size in this picture, are provided with wart-like cytoplasmic projections. **Fig. 52** is a close-up of several superficial cells which resemble strawberries. Here also the cells are covered with small warts.

The protrusion of the superficial cells and the roughness in their cell surface, together with the occurrence of the macroscopic plicae, cause a tremendous increase of the surface area of the synovial membrane. This fact implies an active role of this membrane in the substance transport between the joint fluid and tissue. In fact, there are physiological and morphological evidences of a secretory as well as an absorptive function of the synovial superficial cells.

Note

The synovial membrane of the suprapatellar bursa was taken from the knee of an 18-year-old male in a mid-thigh amputation because of a sarcoma in the tibia. The tissue pieces were fixed in 10 per cent formalin, dehydrated in acetone, dried in air and coated with gold. EM: JSM-2

Reference

See p. 43.

Fig. 52 (×4,800)

12. Rheumatoid Synovial Membrane

As is observed by arthroscopy, rheumatoid arthritis is generally characterized by an unusual growth of synovial villi. The mechanism underlying this change is not fully understood, though it may be explained as a manifestation of a general increase in connective tissue elements, especially of collagen fibers, in this disease. Studies in the sections of the rheumatoid synovial membrane indicate occurrence of lymph follicles and accumulations of plasma cells and macrophages, thus implying the local production of antigens.

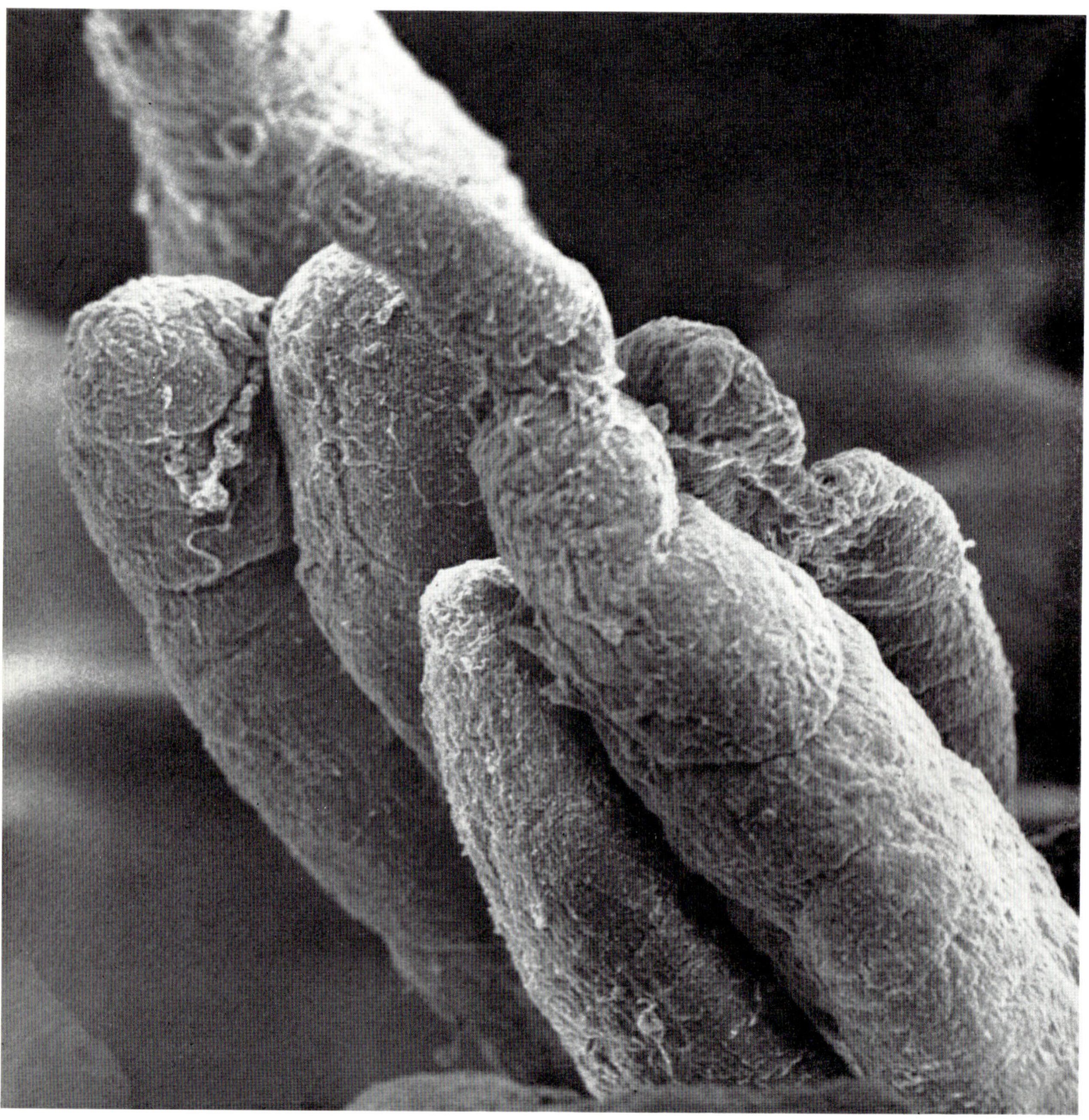

Fig. 53 (× 170)

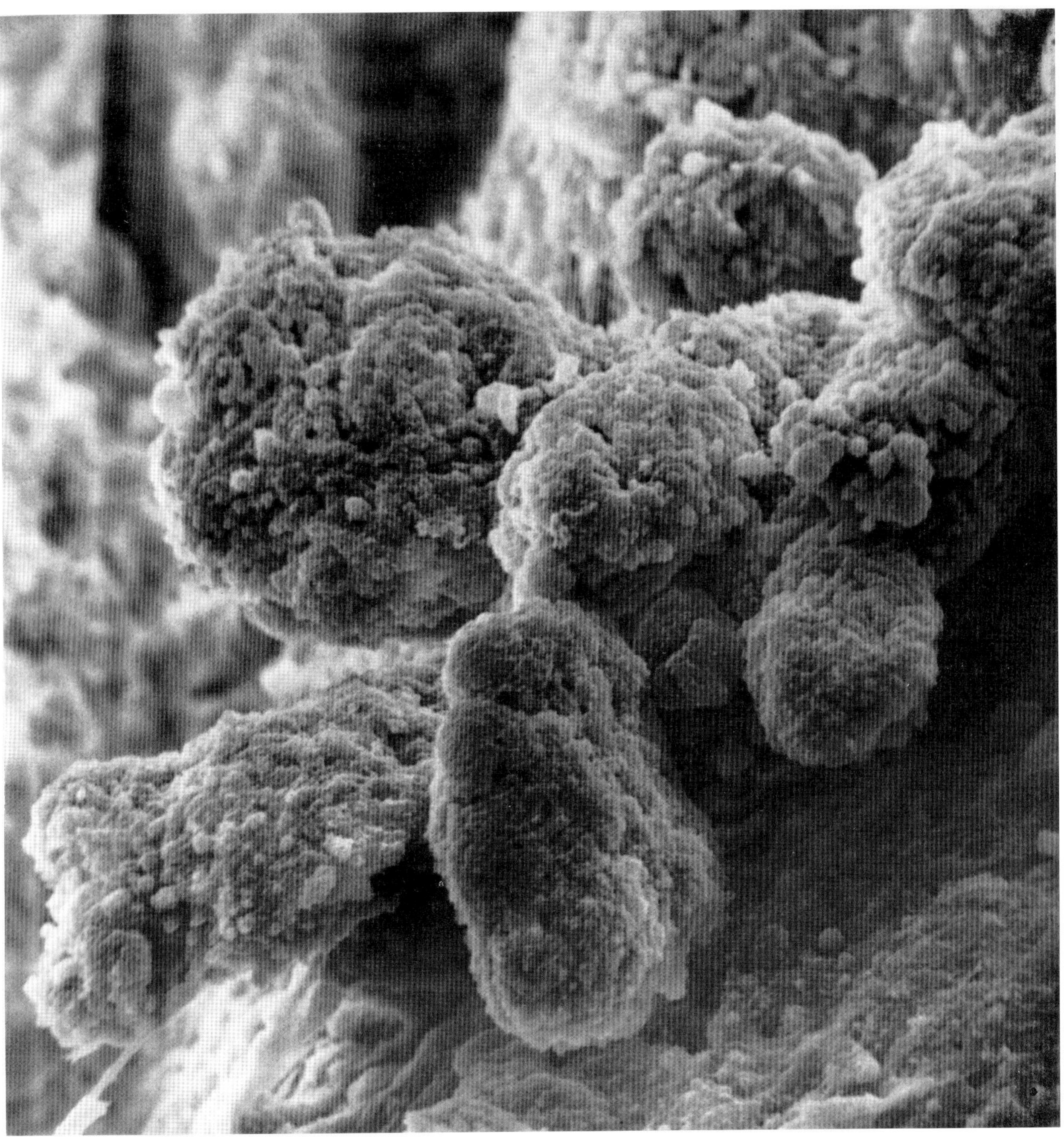

Fig. 54 (× 1,700)

Fig. 53 is a low-power scanning electron micrograph of large, abnormal villi in a rheumatoid joint. Besides such finger-shaped villi, dense, cauliflower-shaped ones may occur. **Fig. 54** shows a view of this category. At this magnification one recognizes some of the superficial cells as round projections.

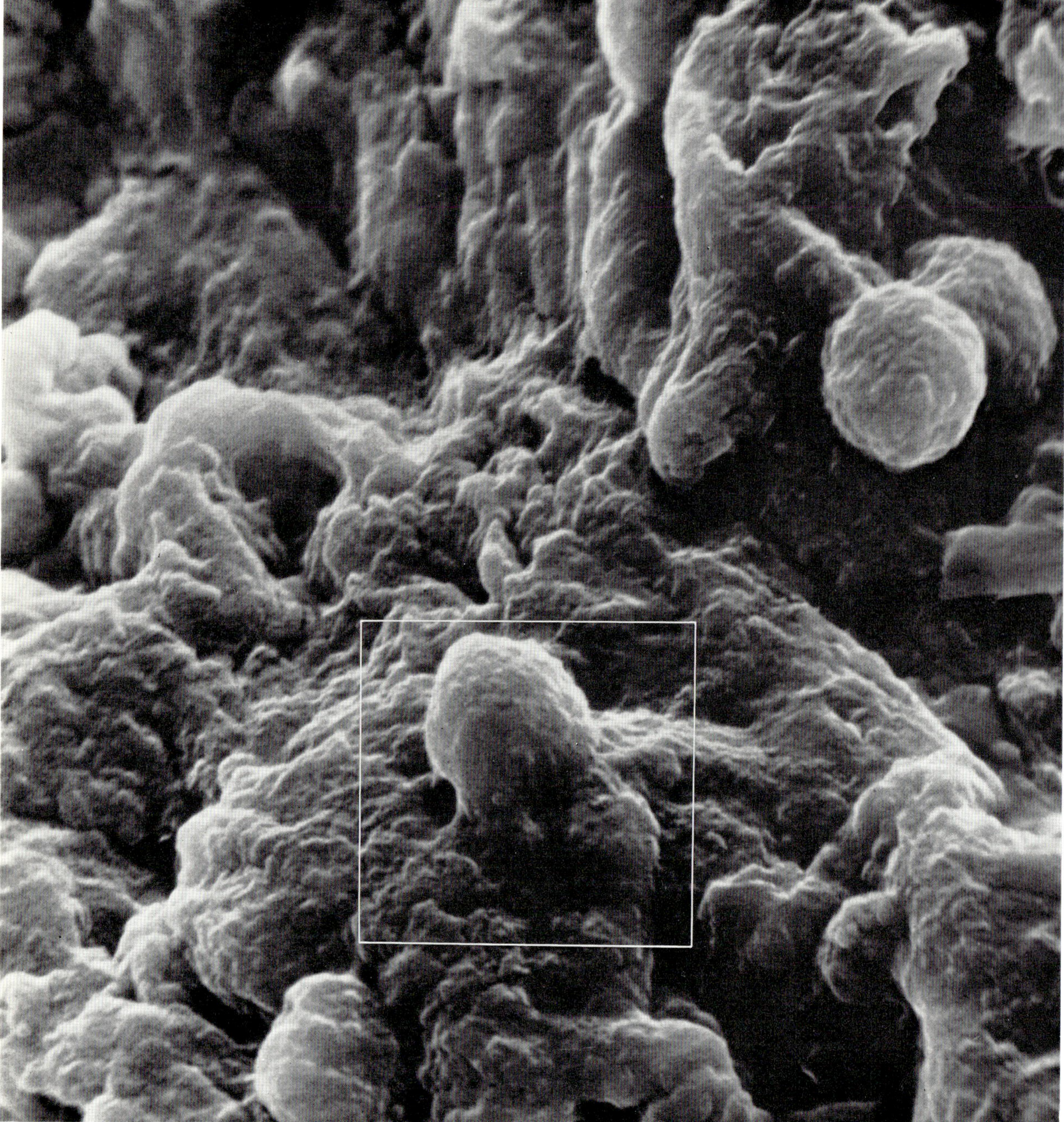

Fig. 55 (×5,100)

The rheumatoid synovial surface may, besides villus formation, be conspicuously uneven with irregular elevations and depressions. As **Fig. 55** shows, a view in a stalactite cave is characteristic of this type of rheumatoid lesion. The polyp-like protrusion of a probable superficial cell (box) is magnified in **Fig. 56**. All the surface appears to be covered by a dull, muddy substance. This substance which mainly consists of the debris of cells and fibers is known to occur abundantly in certain types of rheumatoid arthritis. **Fig. 57** shows this "rheumatoid mud" more distinctly. The round heads of sporadically protruding superficial cells are half buried in this mud.

Fig. 56 (× 12,000)

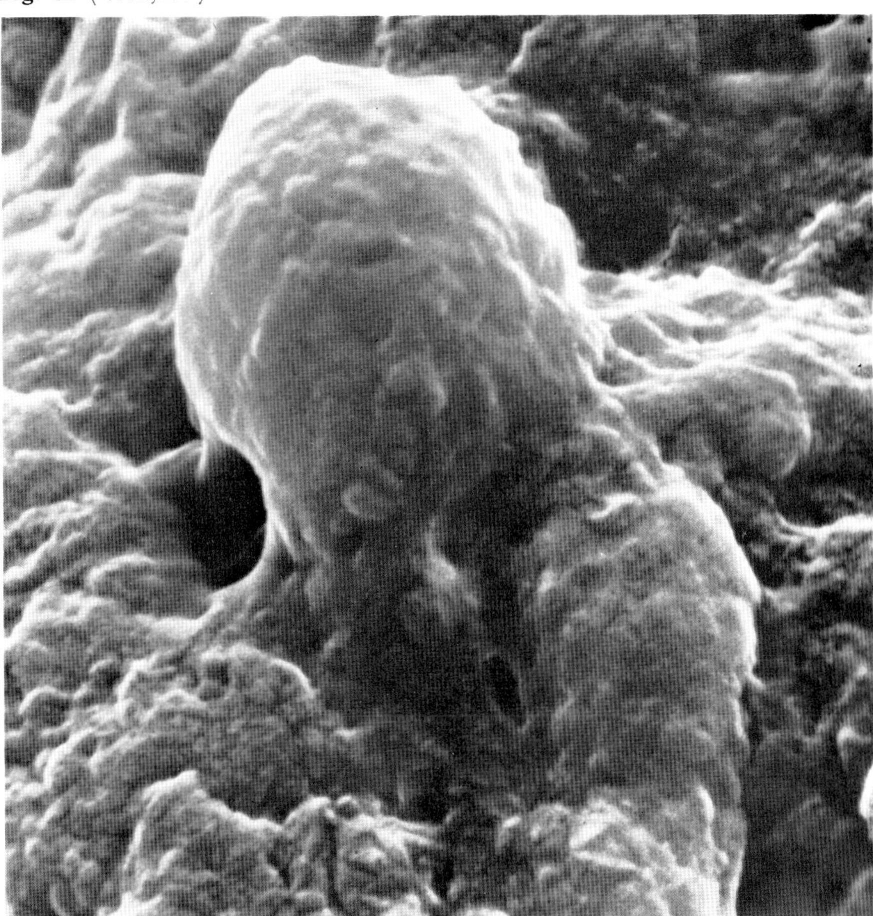

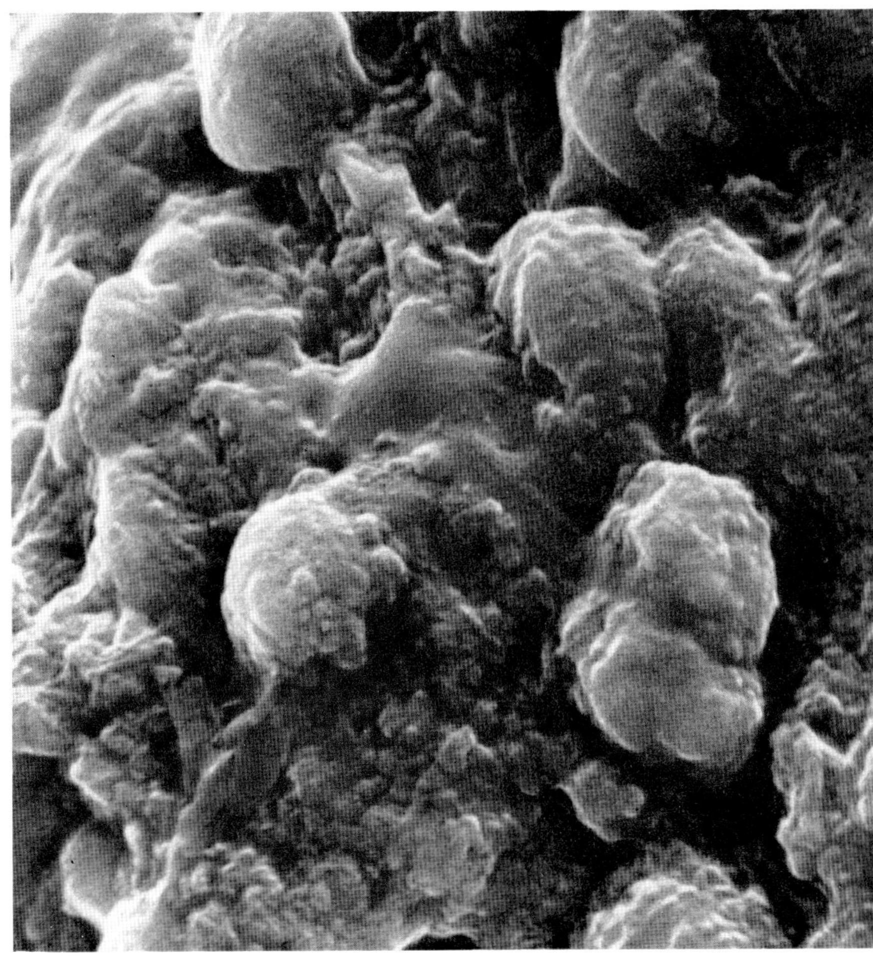

Fig. 57 (× 5,100)

Note

Synovial membranes were obtained in surgical operations from the wrist joint of a 27-year-old female with "definite type" rheumatoid arthritis (Fig. 53), from the elbow joint of a 57-year-old female with "definite type" (Fig. 54), and from the knee joint of a 30-year-old female with "classical type" (Figs. 55–57). After being gently washed with physiological saline, the tissue blocks were fixed in 10 per cent formalin, dehydrated in acetone, dried in air and coated with gold. EM: JSM-2

Reference

FUJITA, T., H. INOUE and T. KODAMA: Scanning electron microscopy of the normal and rheumatoid synovial membranes. Arch. histol. jap. 29: 511-522 (1968).

GRYFE, A., D. L. GARDNER and D. H. WOODWARD: Scanning electron microscopy of normal and inflamed synovial tissue from a rheumatoid patient. Lancet 7612: 156-157 (1969).

Figs. 50–57 by courtesy of the Arch. histol. jap.

13. Meniscus of the Knee Joint

In our knee joint, a pair of consistent plates of fibrocartilage are inserted from the lateral sides of the joint into the articular cavity. These menisci, as they are called because of their crescent form as seen from above, provide, on one hand, additional articular surfaces between the femur and tibia and, on the other hand, serve as elastic cushions to support the body weight and as flexible pads to fill the dead spaces in the joint cavity.

Although the articular surface of the meniscus is as lubric as the surfaces of ordinary joint cartilage, it appears uneven and rough under the scanning electron microscope. This unevenness of the surface is partly caused by the shrinkage of the tissue in the drying procedure, but is partly due to the fibrous forms running along the surface of the tissue.

Fig. 58 ($\times 500$)

Fig. 59 (× 1,200)

As shown in **Fig. 58** groups of parallel fibers, probably collagen fibers, appear on the surface in small areas and hide beneath other fiber groups running in other directions and covering other areas. This interweaving pattern reminds us of the texture of a woolen stuff. Different from the wool fibers in a stuff, the fibers seen in the meniscus surface correspond to the uppermost part of the vertical arcades of collagen fibers. In this superficial part only, the arched fibers run in parallel to the surface.

Figs. 59 and **60** show the central portion of Fig. 58 in closer views. It is obvious that the interwoven fibers are not naked but embedded in a muddy substance which probably corresponds to the polysaccharide-rich ground substance or matrix of the fibrocartilage. Besides, irregular fibers of drop and flame shapes often continuing to fine fibrils lie on the surface apparently free from the embedding substance. They are thought to be collagen fibers and fibrils falling off because of the wear and tear of the articular surface of the meniscus.

Note

Meniscus was taken from the knee of an 18-year-old male in a mid-thigh amputation because of a sarcoma in the tibia. The tissue pieces were fixed in 10 per cent formalin, dehydrated in acetone, dried in air and coated with carbon and gold. EM: JSM-2

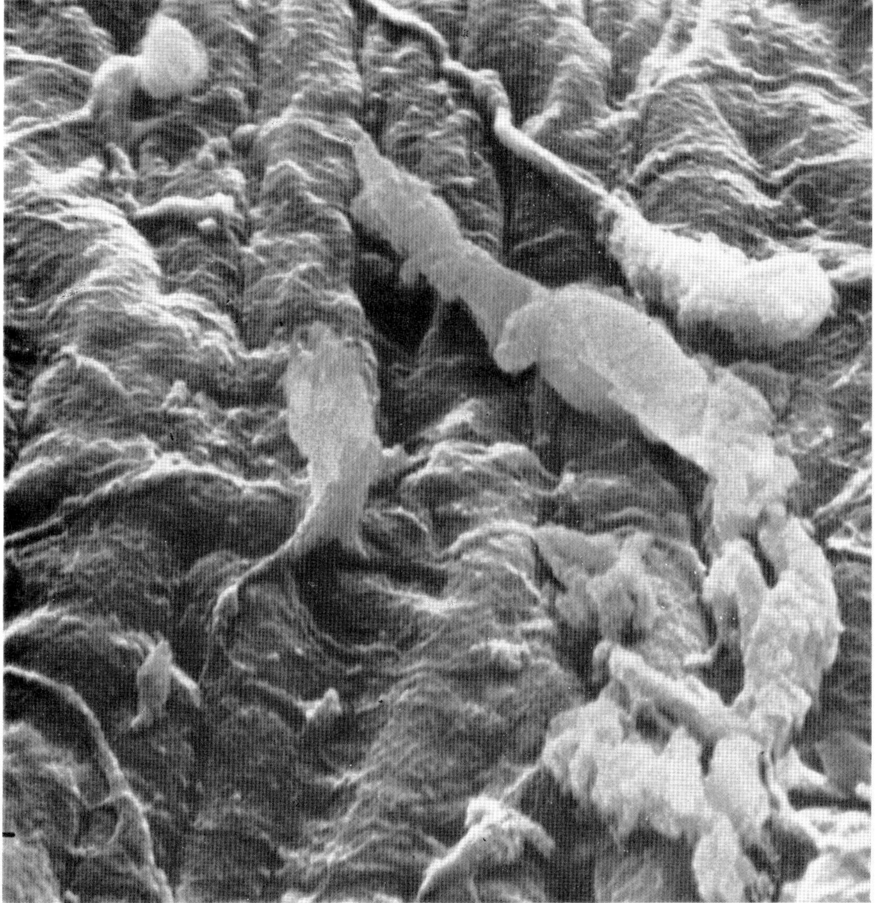

Fig. 60 (× 3,600)

14. Articular Cartilage of the Knee Joint

The ends of bones forming a joint are covered by a few milimeter thick layer of hyaline cartilage. The surface of this cartilage is smooth and covered by highly lubricous synovia or joint fluid. Thus the lubricity in the sliding of two articular surfaces is compared to that of two flat ice surfaces.

When one brings the articular cartilage under the scanning electron microscope, however, its surface is conspicuously rough as thin fibrous structures cover the whole surface (**Fig. 61**). These fibrous structures, which are 0.1–0.3 μ thick and run in rather random directions and in parallel to the surface, likely correspond to small bundles of collagen fibrils. Since Benninghoff (1925) it is known that the collagen fibrils of the articular cartilage form arcades which stand on the bony basis, and that at the top of the arcades the fibrils run in parallel to the surface of the cartilage.

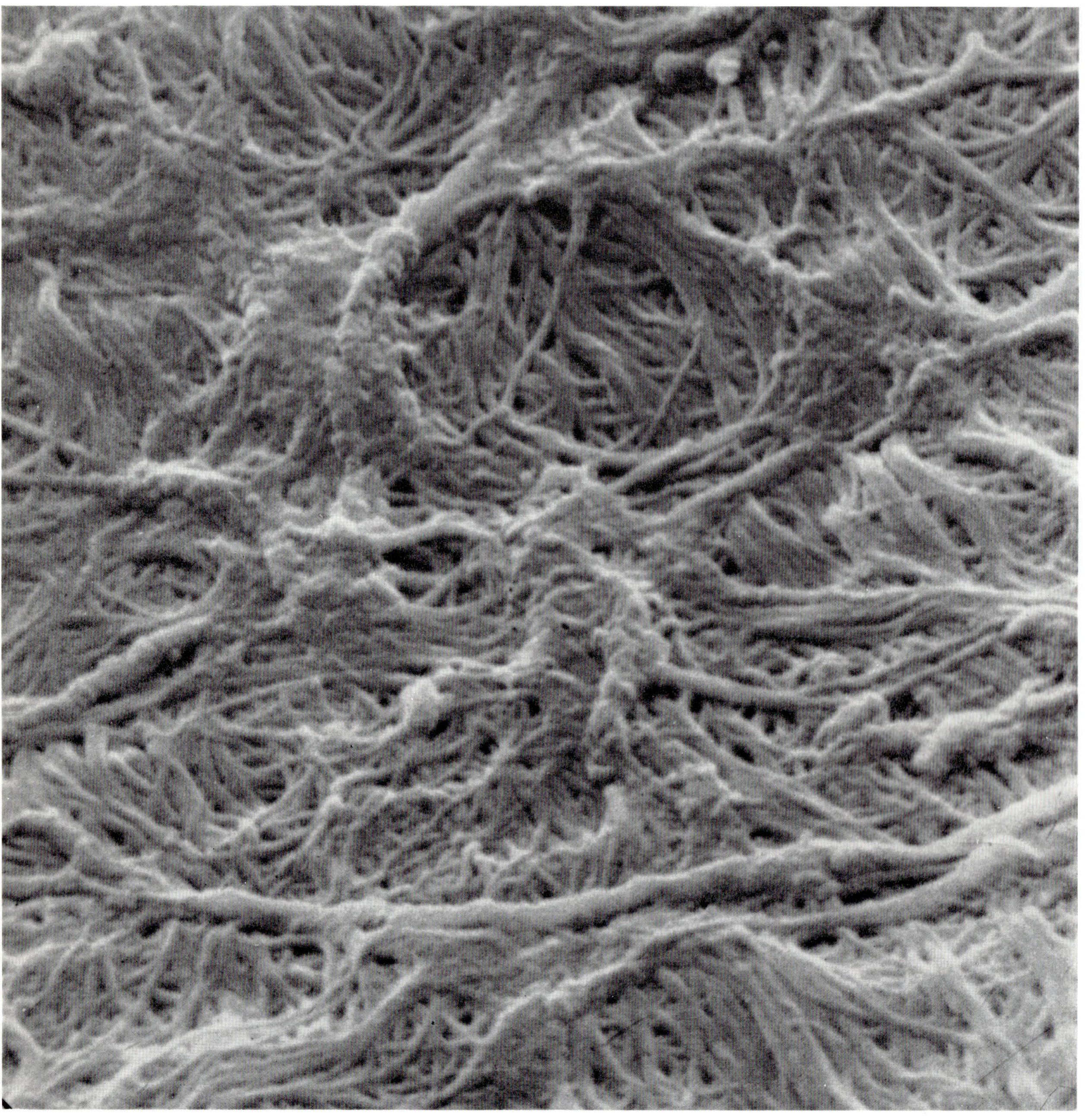

Fig. 61 (×17,000)

Fig. 62 (× 12,000)

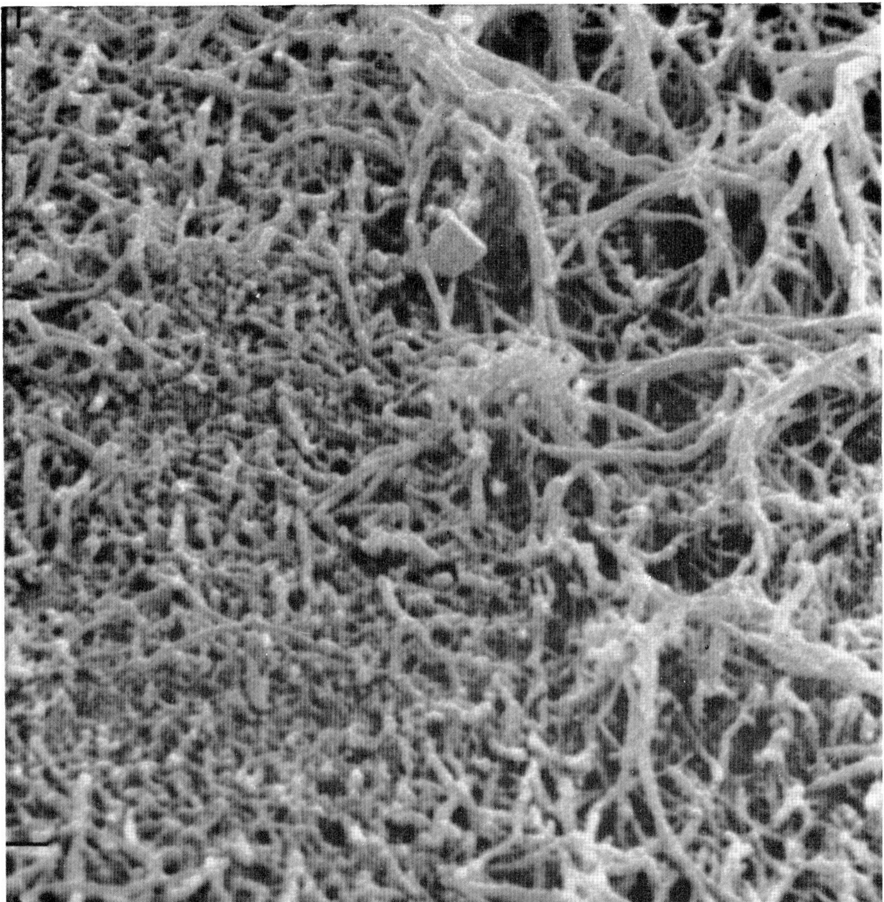

It is noticed that in the scanning electron microscope view of articular cartilage the fibrils appear as if naked and a substance corresponding to the viscous matrix of cartilage is scarcely recognizable.

Figs. 62 and **63** are the interior views of articular cartilage as revealed in its cut surfaces. Small bundles of collagen fibrils are recognized as stubbles. Fig. 63 which corresponds to a layer deeper than that of Fig. 61 indicates cartilage cells and their lacunae, from which the above mentioned fibrils seem to radiate.

Note

Articular cartilage was obtained from the lateral femoral condyle of a 24-year-old female during mid-thigh amputation because of a sarcoma in the tibia. Cartilage pieces were fixed in 10 per cent formalin, dehydrated in acetone and coated with carbon and gold. EM: JSM-2

Reference

GARDNER, D. L. and D. WOODWARD: Scanning electron microscopy and replica studies of articular surfaces of guinea-pig synovial joints. Ann. rheum. Dis. 28: 379-391 (1969).

INOUE, H., T. KODAMA and T. FUJITA: Scanning electron microscopy of normal and rheumatoid articular cartilages. Arch. histol. jap. 30: 425-435 (1969).

MACCALL, J. G.: Scanning electron microscopy of articular surfaces. Lancet 2: 1194 (1968).

WALKER, P. S., J. SIKORSKI, D. DOWSON, M. D. LONGFIELD, V. WRIGHT and T. BUCKLEY: Behaviour of synovial fluid on surfaces of articular cartilage. A scanning electron microscope study. Ann. rheum. Dis. 28: 1-14 (1969).

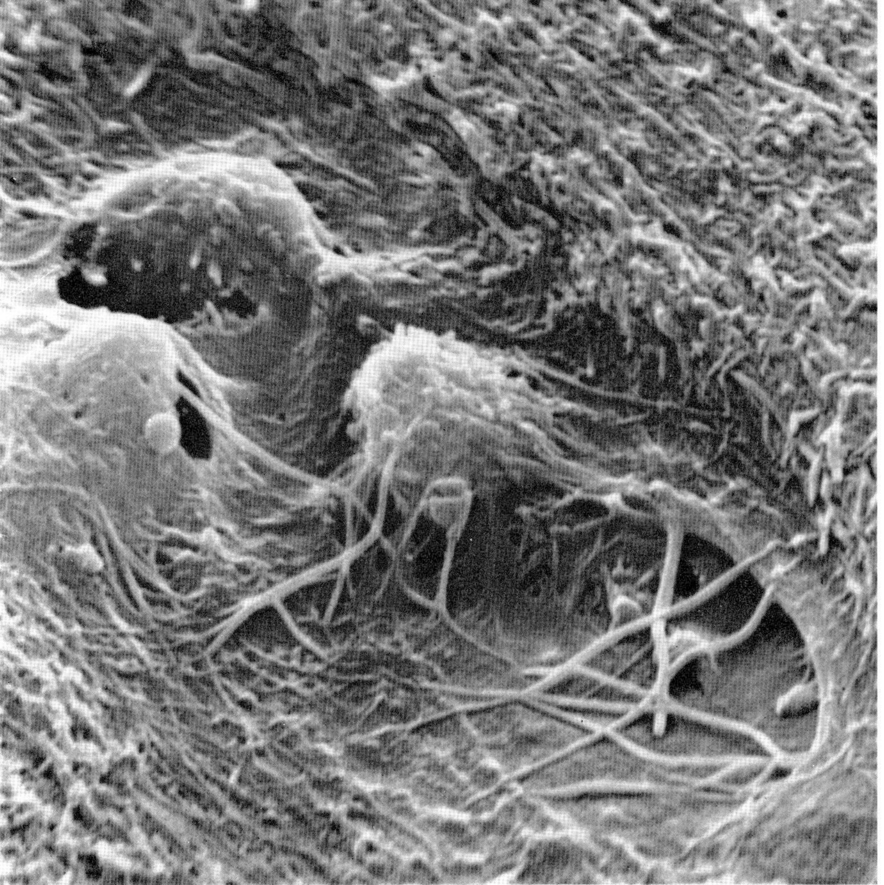

Fig. 63 (× 7,200)

15. Rheumatoid Articular Cartilage

The changes in the articular surfaces caused by rheumatoid arthritis have been studied with difficulties by previous methodologies but can easily be investigated by the use of scanning electron microscopy.

Figs. 64 and **65** show at different magnifications the articular surface from a rheumatoid knee joint of an old patient. The arrangement of the collagen fibrils on the surface is markedly disturbed and larger and smaller cavities are formed undermining the superficial fibrils.

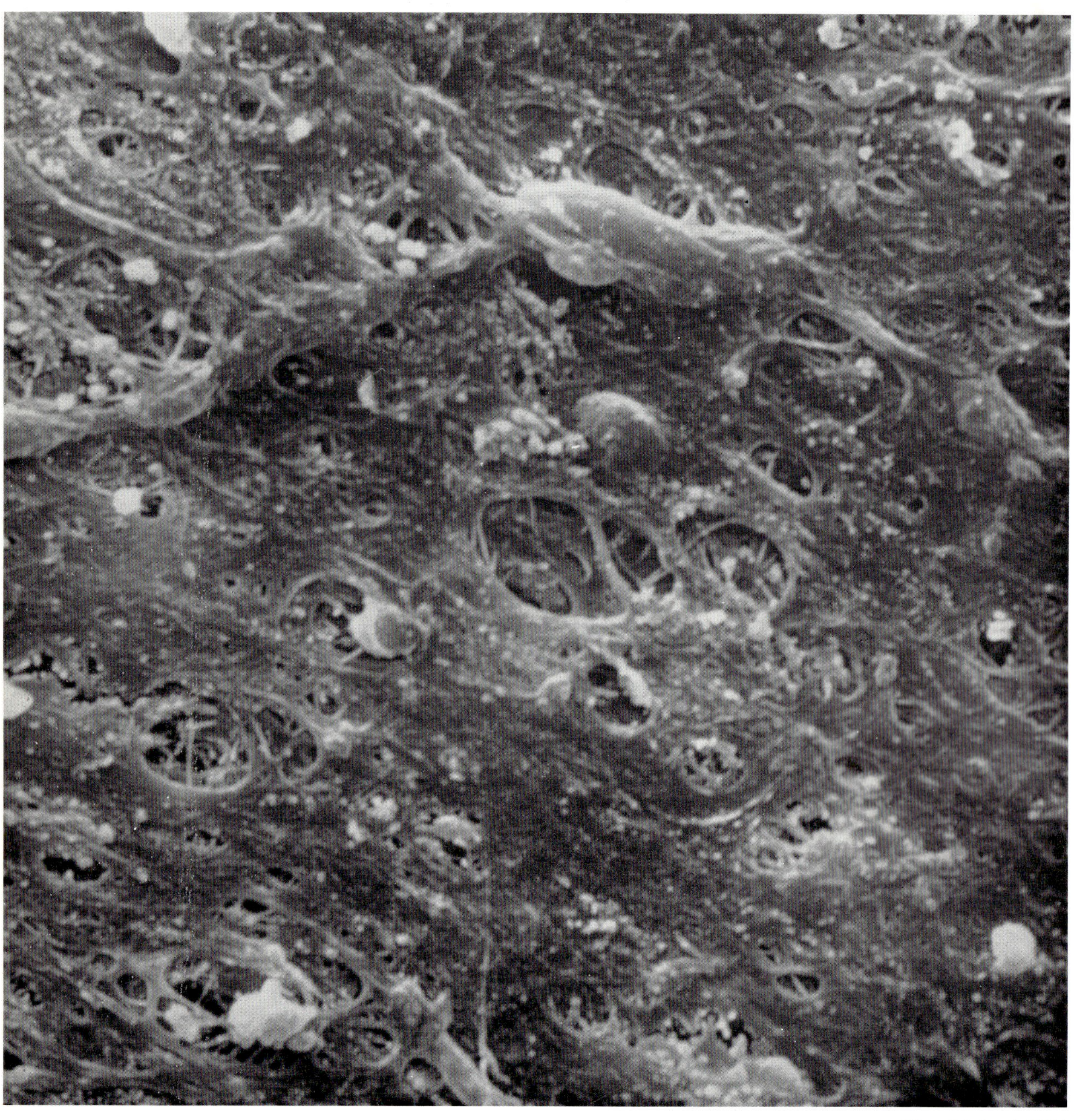

Fig. 64 (× 1,700)

Fig. 65 (× 3,600)

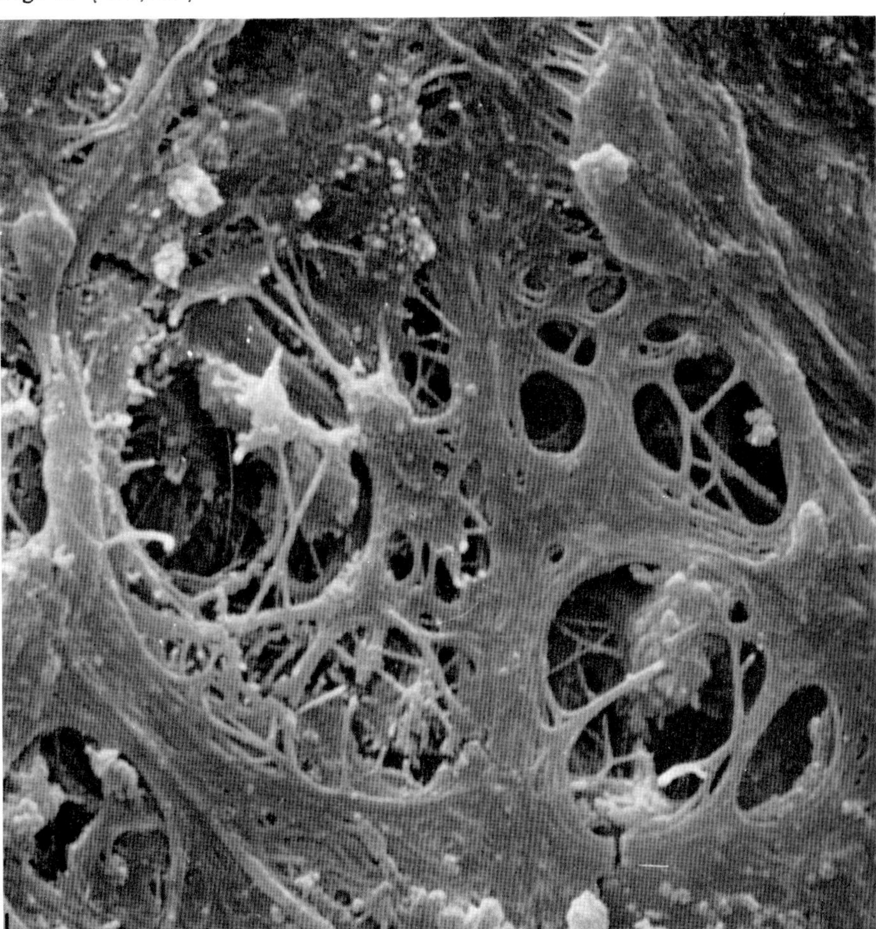

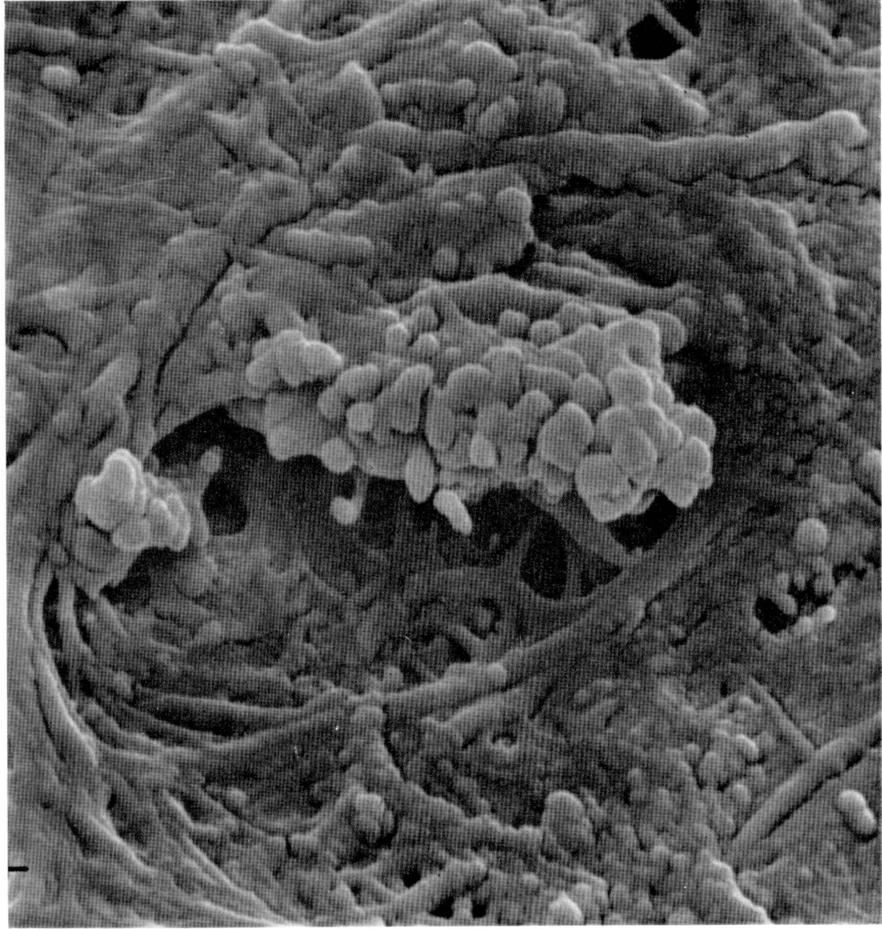

Fig. 66 (× 12,000)

Fig. 66 also indicates one of such cavities in the articular cartilage from a rheumatoid elbow joint of a young woman. Noteworthy is the occurrence of abundant round granules on the surface of fibrilar structures. Some granules are gathered like a bunch of grapes. It is not known whether these granules represent a degenerated form of collagenous fibrils or a substance of other sources attached to the fibrils.

Note

Rheumatoid articular cartilages were taken from the lateral condyle of the femur of a 57-year-old female during the debridement of the knee joint, and from the head of the radius of a 24-year-old female during the arthroplasty of the elbow joint. Specimen preparation as in normal cartilages (p. 47). EM: JSM-2

Reference

See p. 47

Figs. 64-66 by courtesy of the Arch. histol. jap.

SKIN AND SENSORY ORGANS

16. Surface of the Human Skin

The superficial layer of the skin is the epidermis which consists of polyhedral cells piled up like a stone wall. The epidermal cells are generated in the basal layer of the epidermis and, moving upward, become cornified in the superficial layer to be desquamated.

Fig. 67 is a scanning electron micrograph of the hip skin of a girl. One may be surprised at how rough and rude her skin surface really is when it appears so soft and

Fig. 67 (× 1,700)

Fig. **68** (× 3,600)

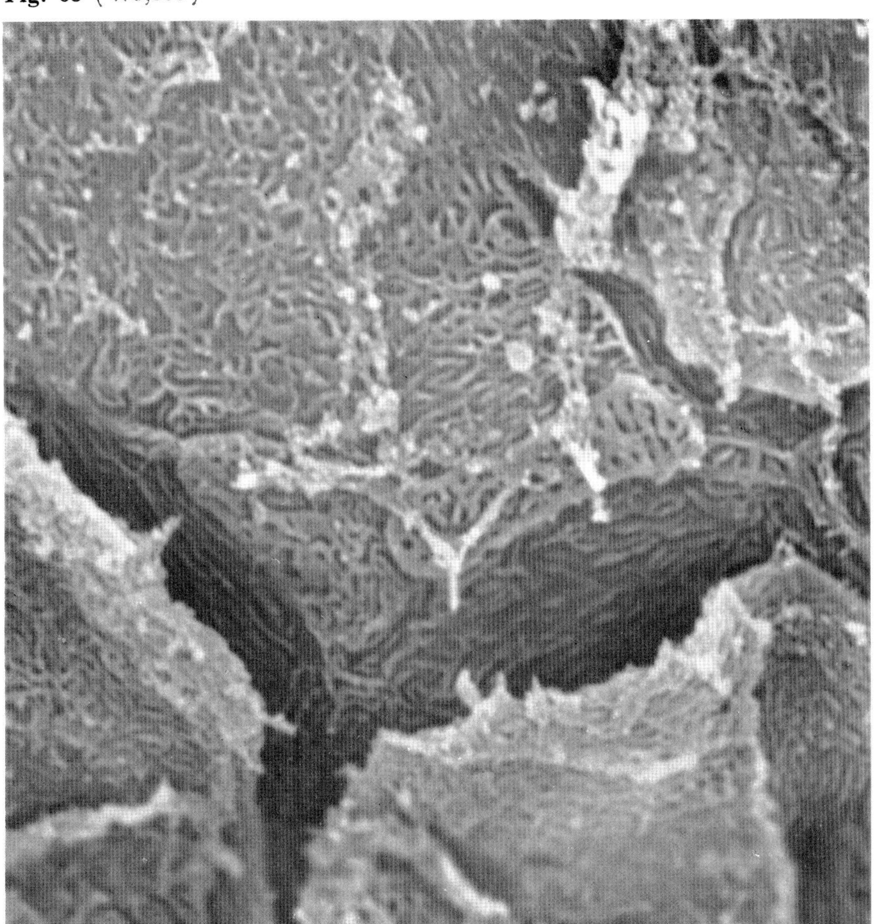

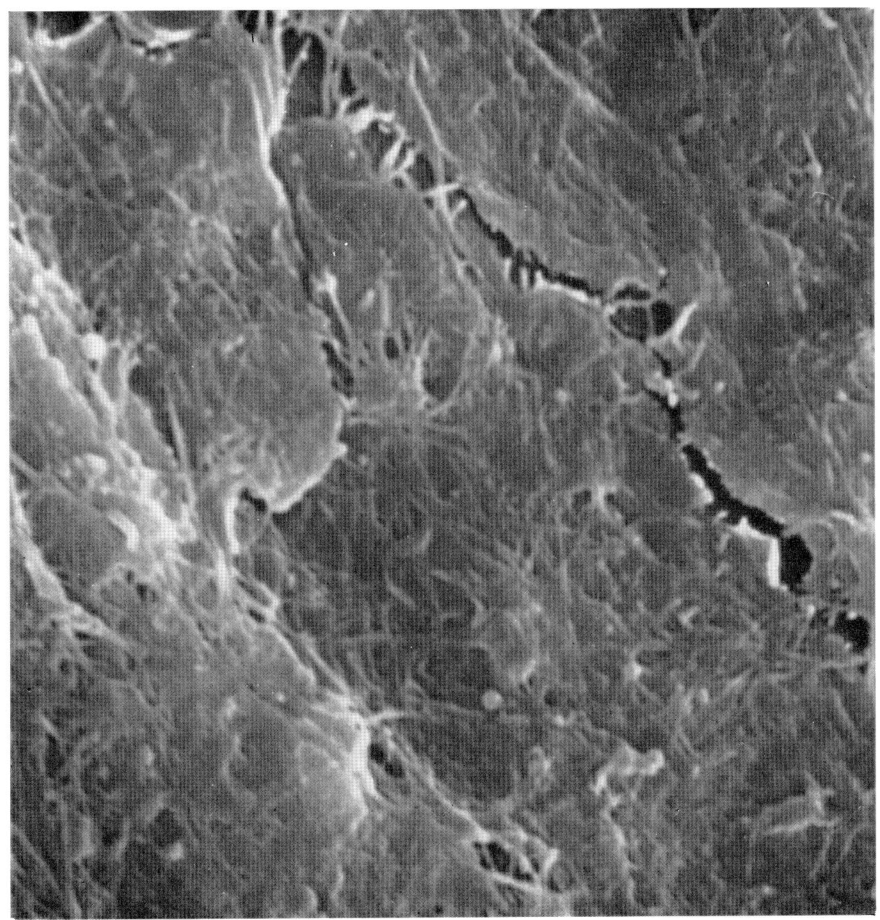

Fig. 69 (× 3,600)

smooth to the naked eye. Flattened and cornified cells, mainly hexagonal in shape, loosely pave the sur face; some are raised as if just peeling off. **Fig. 68** shows a few cornified cells in a closer view. A characteristic fine relief is recognized on the cell surface.

The surface aspect of the epidermal cell may vary in different dermatological conditions. As a conspicuous example, the epidermis covering a scar is shown in **Fig. 69** at the same magnification as in Fig. 68. The fibrous elements running through the cytoplasm and over the intercellular clefts correspond without doubt to the tonofilament bundles and intercellular bridges found in studies of sections of the tissue.

Note

Pieces of the skin obtained from a post-operative scar and from an adjacent normal part on the hip of an 18-year-old female were fixed in 10 per cent formalin, dehydrated in acetone, dried in air and coated with carbon and gold. EM: JSM-2

17. Collagen and Lattice Fibers of the Human Dermis

The dermis or corium of the skin is a layer directly beneath the epidermis. It is a compact feltwork of collagen fibers which are enormously resistant to tension. The dermis of animals is utilized as leathers.

The fibers in the deeper portion (reticular layer) of the human dermis are shown in **Fig. 70.** The thick, wavy fibers are collagen fibers. Every fiber is recognized as a bundle of numerous filaments called collagen fibrils. Some of the fibrils, however, are not involved in the formation of the collagen fibers but form a delicate network. These loose fibrils are called reticular or lattice fibers.

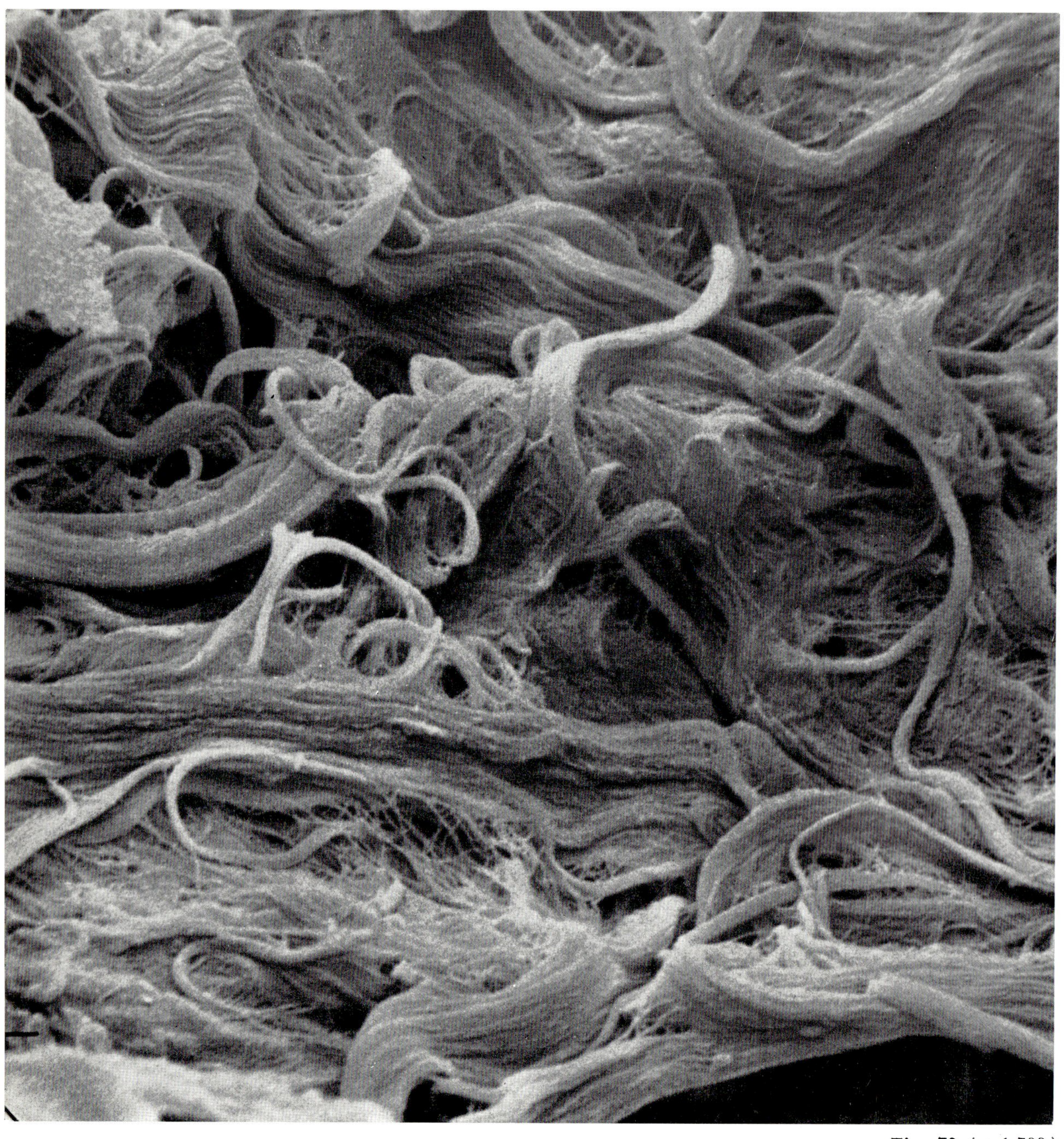

Fig. 70 (×1,700)

Fig. 71 (×3,600)

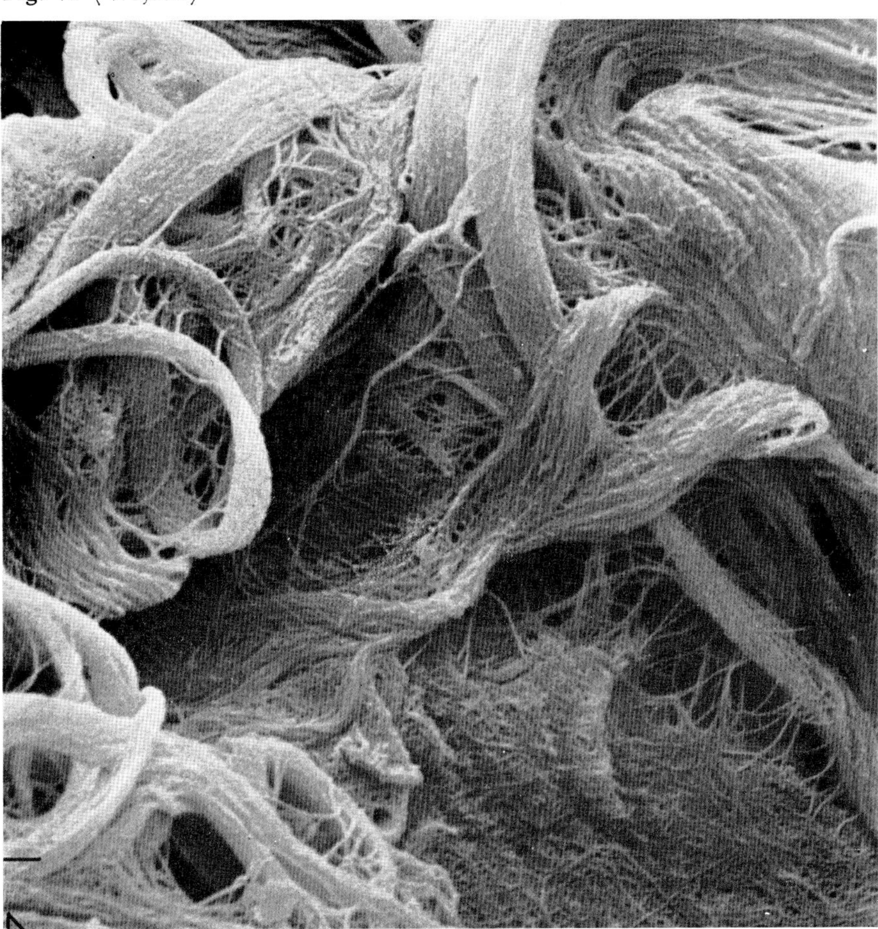

An interfibrous substance and cellular elements which should have filled the spaces among the fibers seem to have been lost when the cut surface of the tissue was washed with a hypotonic buffer.

Figs. 71 and **72**, which are closer views of an area of Fig. 70, show a continuous relation of collagen and lattice fibers. Fibrils bundled up into a collagen fiber become loosened and gradually become part of the network of fibrils.

The banding relief of the collagen fibrils which is obvious in transmission electron microscopy has not yet been recognized by scanning electron microscopy, presumably because the metal coating hides the fine surface structure of the fibrils.

Note

The skin of the head of a 60-year-old male was excised. Vertical cut surfaces were made and thoroughly washed with a 0.1M phosphate buffer (pH 7.2). The specimens were then fixed in 10 per cent formalin, dehydrated in acetone, dried in air and coated with carbon and gold. EM: JSM-2

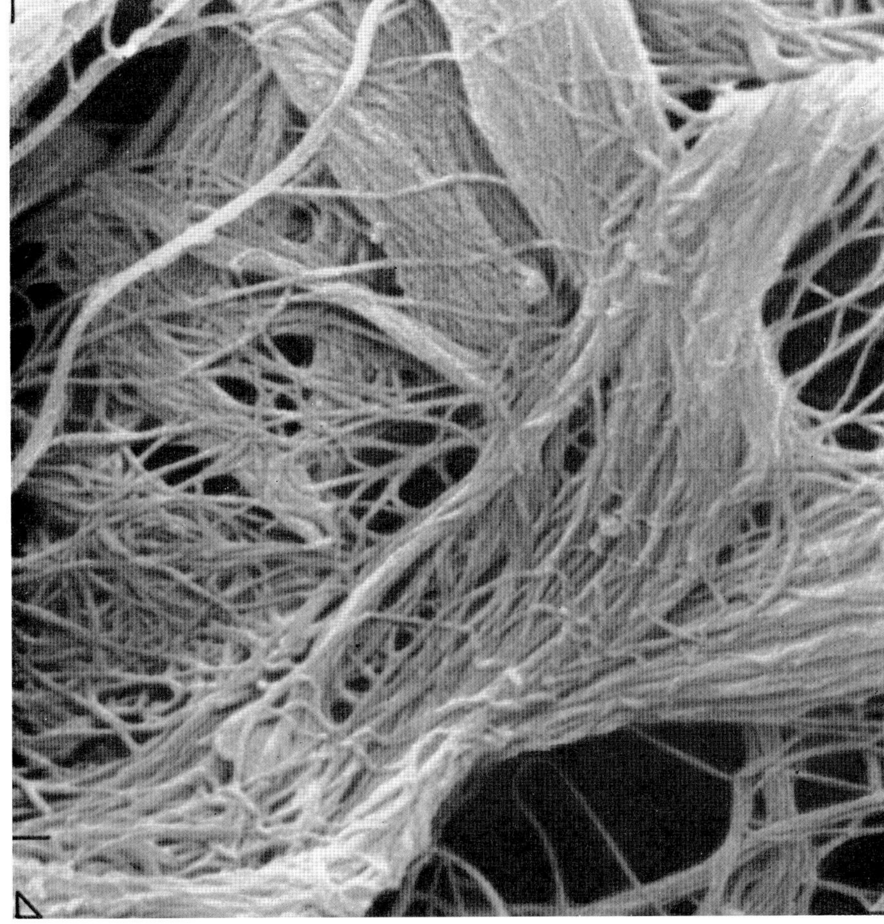

Fig. 72 (×12,000)

18. Human Hairs

The hairs are keratinous filaments projecting from a deep invagination of the epidermis of the skin. The hairs grow at the bottom of this invagination or hair follicle.

Fig. 73 shows the basal part of a hair of the human head. The epidermal surface is covered with leaf-like cells which, when peeled off, will form dandruff. The deep groove of the epidermis surrounding the hair base corresponds to the orifice of the hair follicle. Sebaceous or oily secretion moistening the hair also comes out from this groove. The hair itself is covered with scale-like plates called epicuticles which are an extremely flattened modification of keratinized epidermal cells.

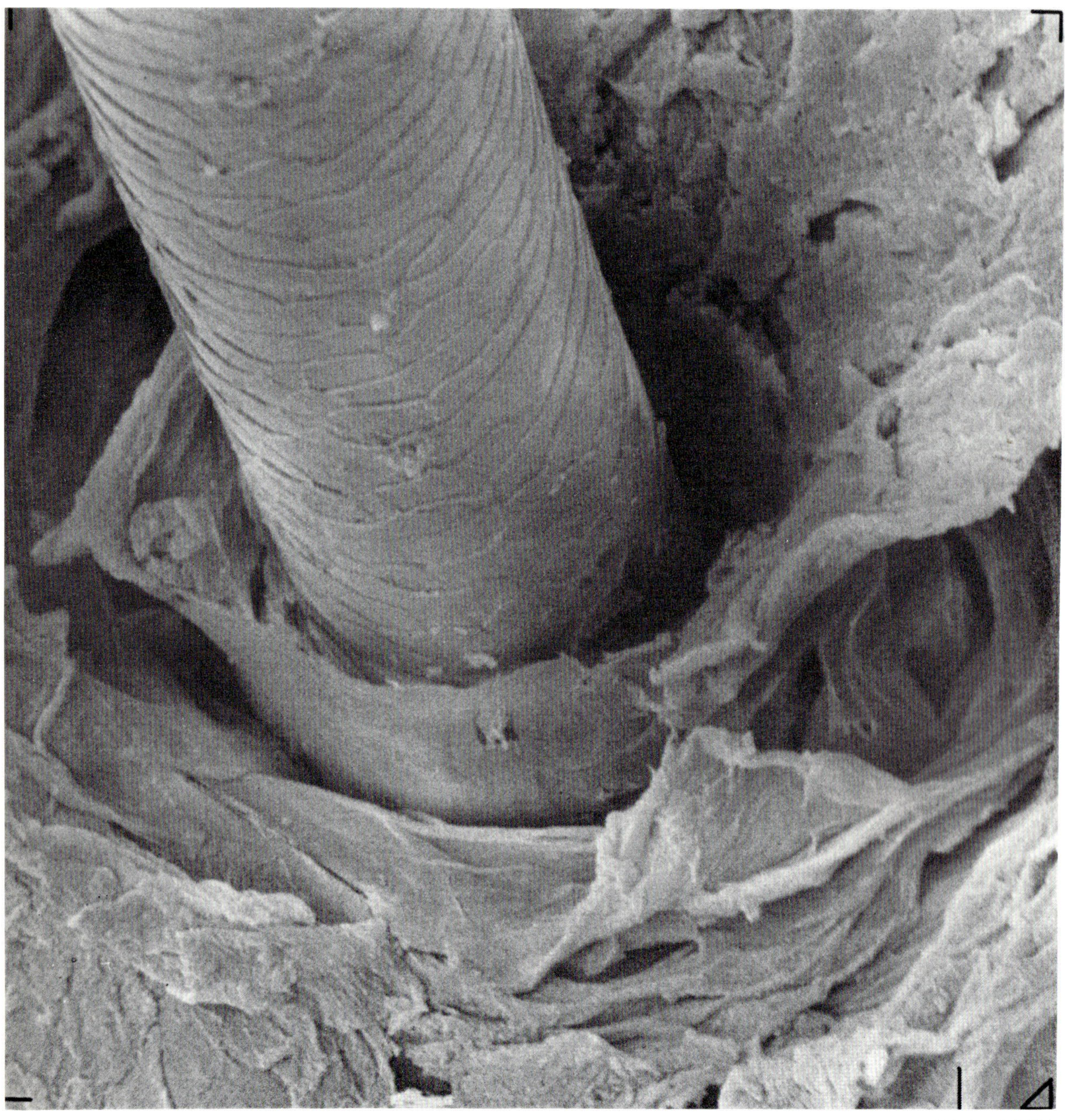

Fig. 73 (× 1,700)

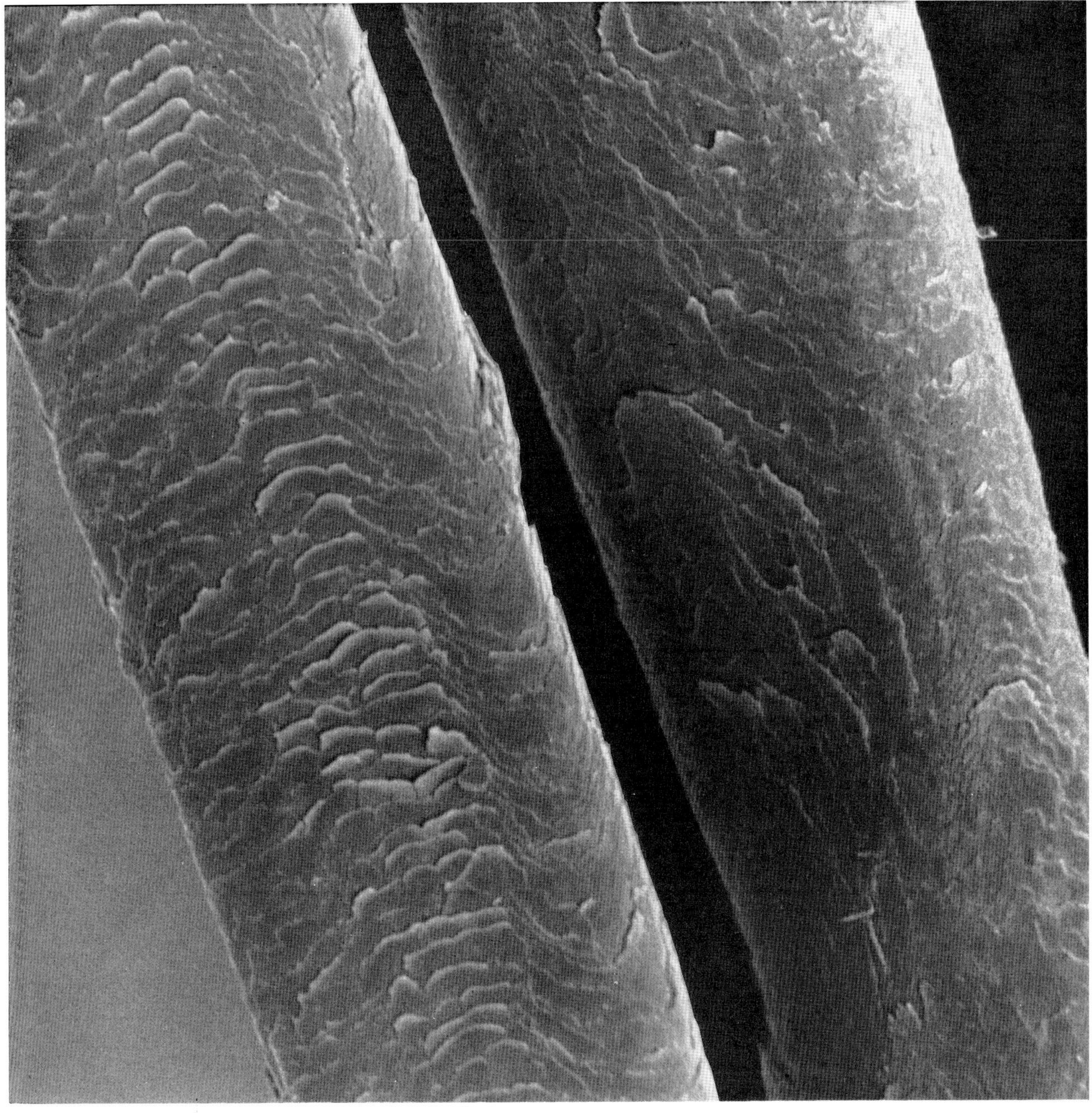

Fig. 74 (×510)

Fig. 74 shows a hair of the eyebrow (left) and a pubic hair (right) from a 30-year-old man. A conspicuous difference in the surface structure of these hairs is recognized. In the head hair (Fig. 73) the lines of the epicuticle borders run transversely and at regular intervals. The hair of the eyebrow shows a less regular pattern of the border lines of epicuticles, whereas in the pubic hair the lines are strongly twisted and very irregularly arranged.

Fig. 75 is a closer view of an axillary hair taken from the same 30-year-old man. Serrated borders and longitudinal striations of the epicuticles are obvious.

Figs. 76 and **77** are the surface views of a hair of a 23-year-old woman permanently waved two months previously. The epicuticles are seen eroded by the strong chemical treatment in permanent waving.

Previous transmission electron microscope studies using replicas of hairs have already indicated that the pattern of the epicuticle coverage differs not only among different kinds of hairs but also in the nature of the hairs, whether tough, smooth or curled (Hirai and Yamato, 1957; Tsuda and Nishi, 1960). Scanning electron microscopy is expected to be useful for individual identification as well as racial comparison of hairs.

Fig. 75 (× 1,200)

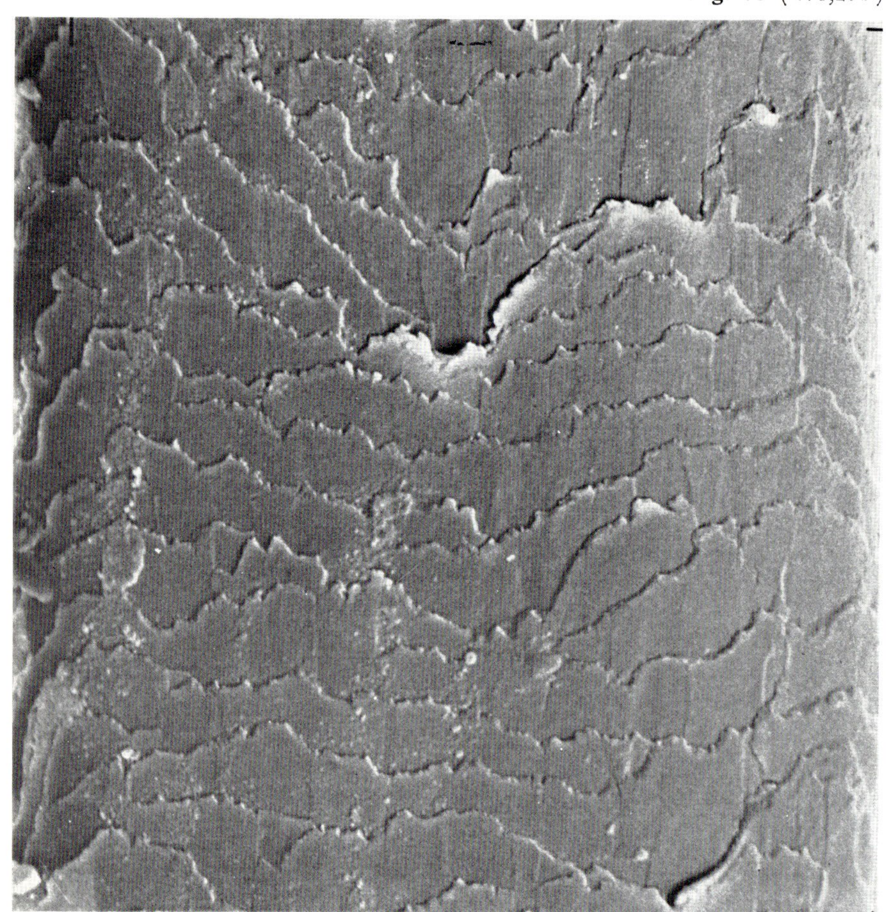

Note
The material shown in Fig. 73 was taken from the cadaver of a 65-year-old male. This specimen and the hairs shown in other figures were fixed in 10 per cent formalin. The specimens were dehydrated in acetone, dried in air and coated with carbon and gold. EM: JSM-2

Reference See p. 59.

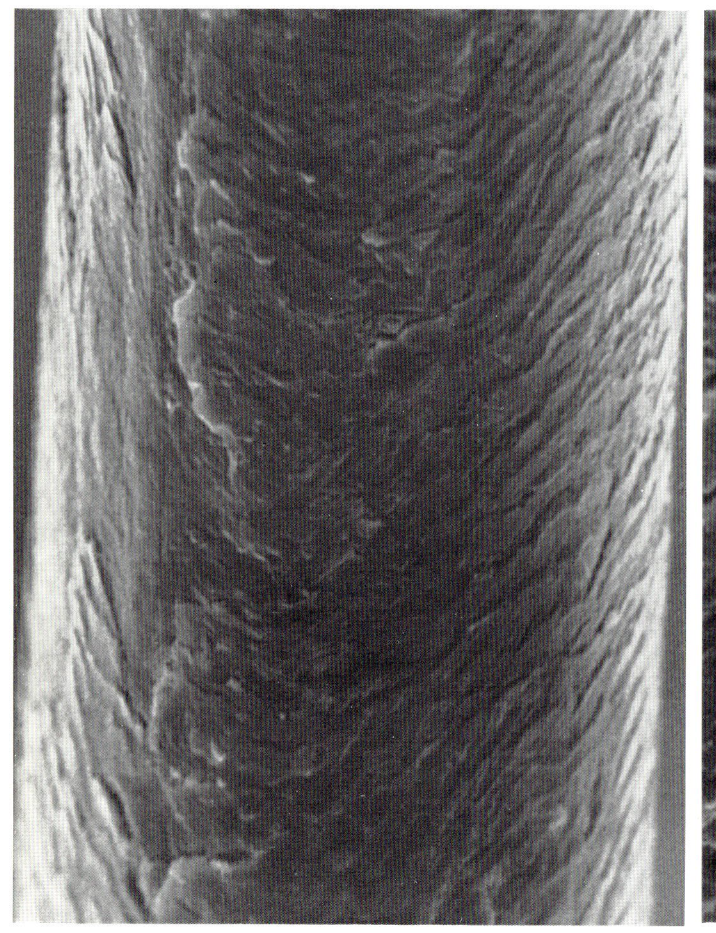

Fig. 76 (× 1,200)

Fig. 77 (× 3,600)

19. Hairs of Animals

The body hairs of mammals show species differences in their surface fine structure as has been shown in some species by use of the replica method (HIRAI and YAMATO, 1957).

Fig. 78 shows the hairs of a mouse. Characteristic are the wide-spaced lines of epicuticle borders running transversely or obliquely. As indicated by an arrow, a longitudinal furrow often occurs in the hairs of this animal.

Fig. 78 (×1,700)

Fig. 79 (×3,600)

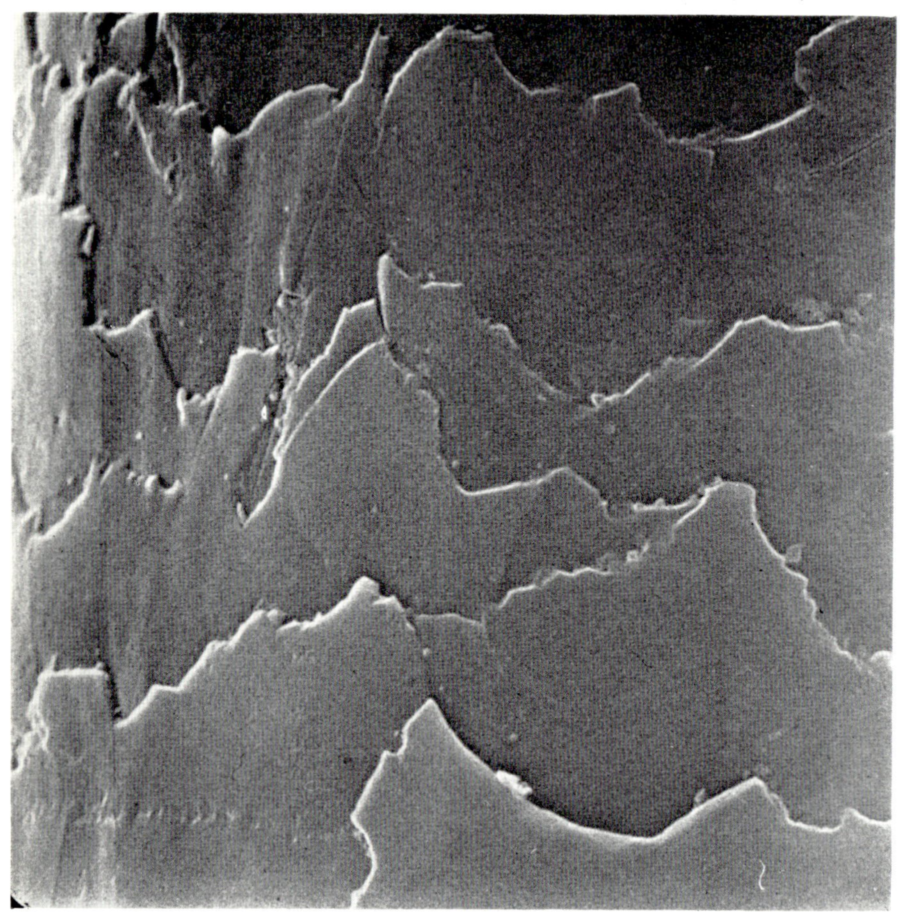

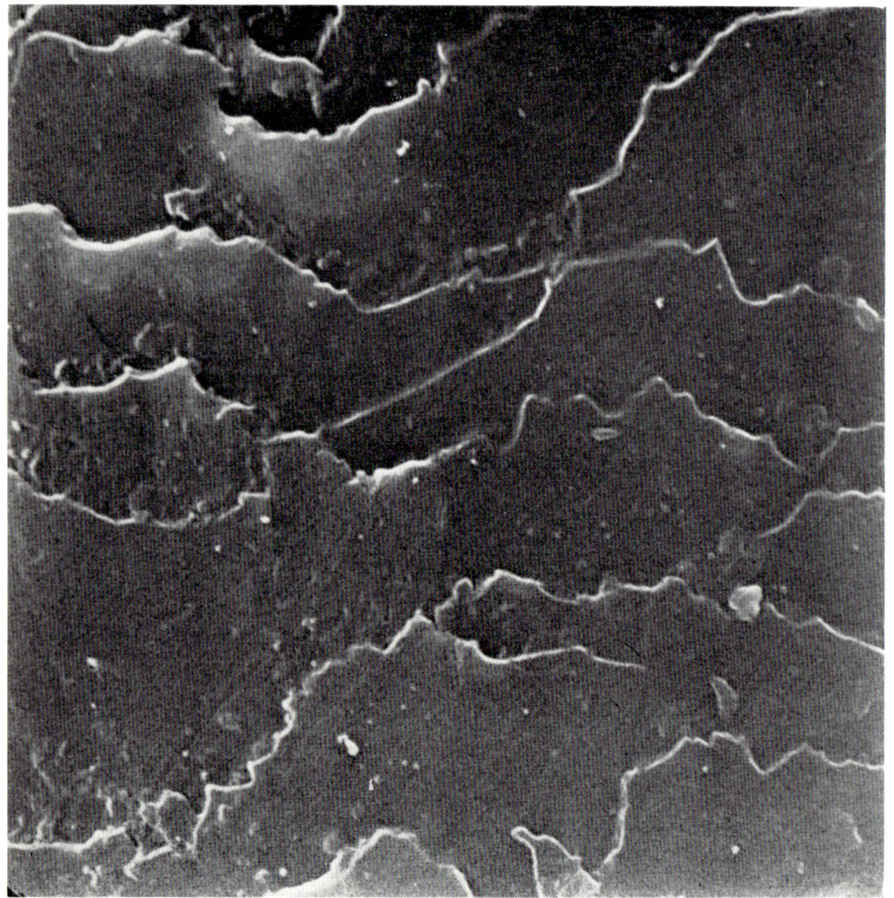

Fig. 80 (×3,600)

Figs. 79 and **80** are the surface views of donkey and rhesus monkey hairs, respectively. These hairs, with very thin, slate-like epicuticles, resemble human hairs. The scale borders have a zigzag pattern.

Figs. 81 and **82** are high-power micrographs of the hairs of the camel and tiger. The camel hair (Fig. 81) shows relatively smooth arches of scale borders. In the tiger hair (Fig. 82), the epicuticles are thick and have a marked longitudinal striation. Some scales are polyhedral in form and reveal sharp edges of a Y-shape.

Note

The hairs shown in Fig. 78 were fixed in 10 per cent formalin while all others were not fixed. Carefully washed materials were dehydrated in acetone, dried in air and coated with carbon and gold. EM: JSM-2

Fig. 81 (× 12,000)

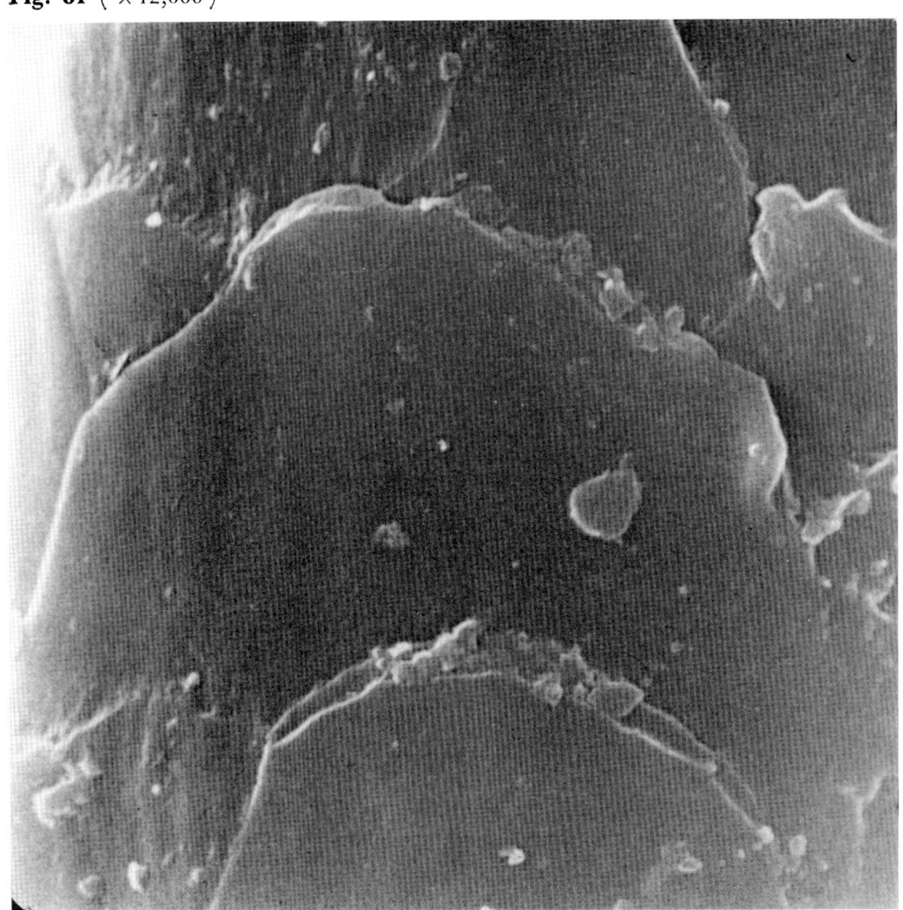

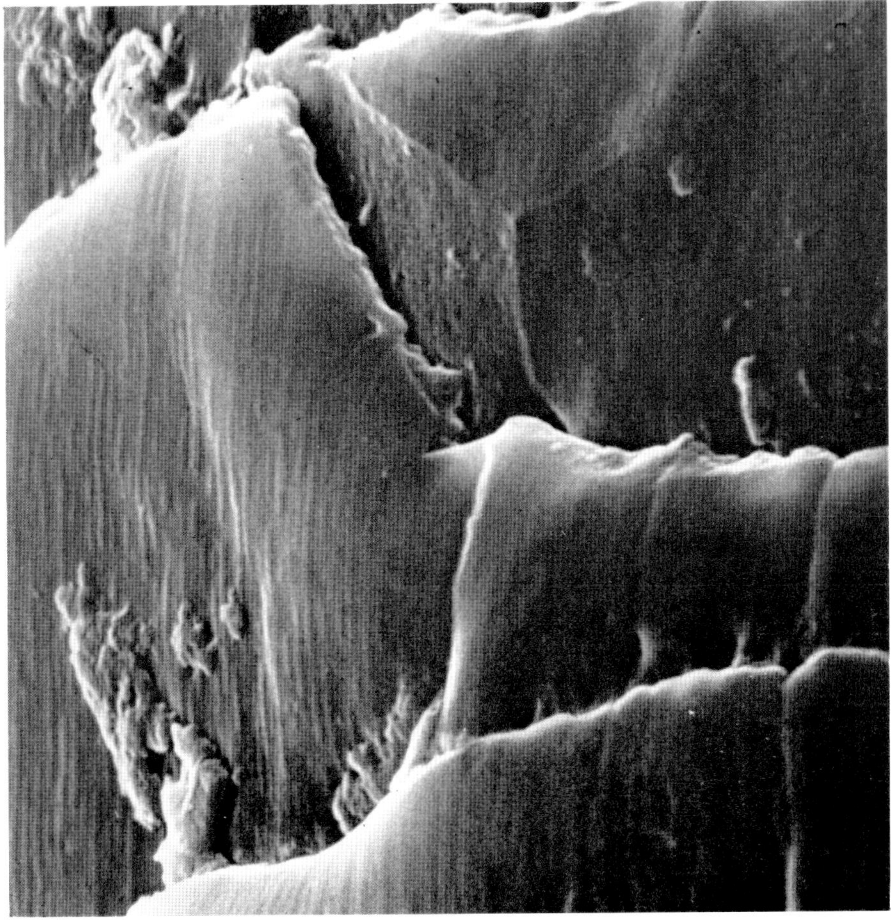

Fig. 82 (× 12,000)

Reference

GOLDMAN, L., J. VAHL, R.J. ROCKWELL, R. MEYER, M. FRANZEN, P. OWENS and S. HYATT: Replica microscopy and scanning electron microscopy of laser impacts on the skin. J. invest. Dermatol. 52: 18-24 (1969).

HIRAI, Y. and K. YAMATO: Electron microscopic studies on the surface structure of the human and animal hairs (in Japanese). Juzen Igakukai Zasshi 59: 1085-1088 (1957).

RAMANATHAN, N., J. SIKORSKI and H.J. WOODS: Electron microscope studies of the surface structure of wool and other fibres. Biochem. biophys. Acta 18: 323-340 (1955).

SUDO, T.: Nippon Atlas of Hairs (in Japanese). Nippon Mohatsu Kagaku Kyokai, Tokyo, 1966.

TSUDA, H. and M. NISHI: Electron microscopy of the epicuticle pattern of the human hair, as shown by replica preparation (in Japanese). Acta anat. nippon. 35: 140-151 (1960).

20. Celluloid Impression of the Human Skin

When a celluloid plate is softened by a drop of amyl acetate and pressed for a few minutes on a given matter, the fine unevenness on the surface of the latter is clearly reproduced on the celluloid plate which can then be observed under the microscope. Since its invention by SUZUKI in 1932, this SUMP method, so-called from the initials of SUZUKI's Universal Micro-Printing, has contributed much to various fields, especially to the studies of textiles, rocks, plants and animals. The technique is so simple that it has also been used widely in the education of school children in Japan.

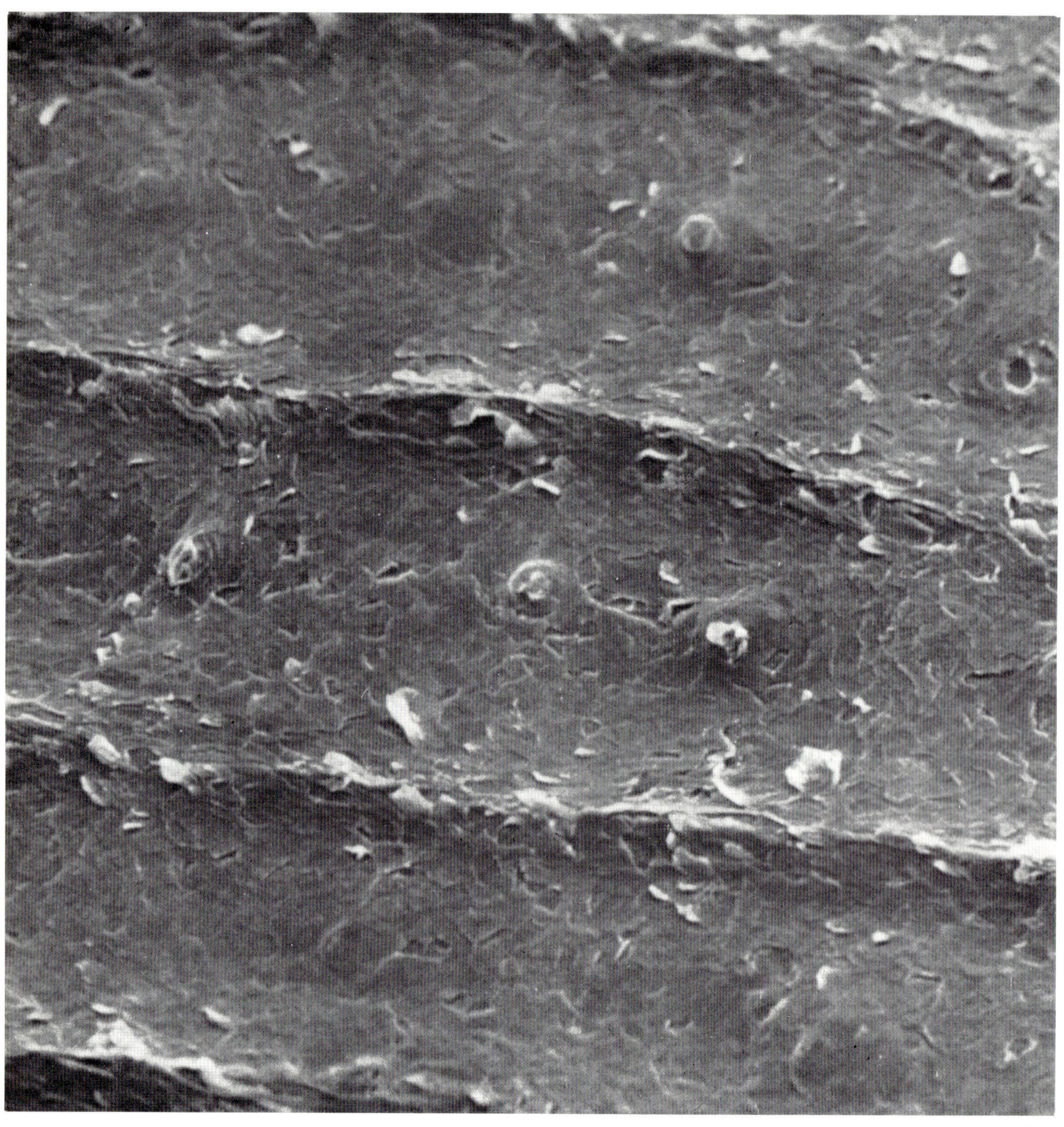

Fig. 83 (×170)

Fig. 84 (× 170)

The SUMP skin replicas, when coated with metals, have been shown by the authors (FUJITA et al., 1969) to be fully applicable to scanning electron microscope observation. They are free from deformation and shrinkage which would be unavoidable if the skin itself were fixed, dried and brought under the scanning electron microscope.

Figs. 83, 85 and **86** are low and high power scanning electron micrographs of a SUMP replica from the finger belly. In **Figs. 84, 87** and **88** the photographs are printed negatively so that the black-and-white relation appears in reverse to that in Figs. 83, 85 and 86. The natural stereological aspects of the skin surface seem to have been approximately reproduced by photographic magic.

Figs. 83 and **84** show the crests and furrows of the finger print. The scale-like cells covering the skin are already visible at this magnification. On the middle of the crests the ducts of sweat glands open (arrows).

Fig. 85 (×360)

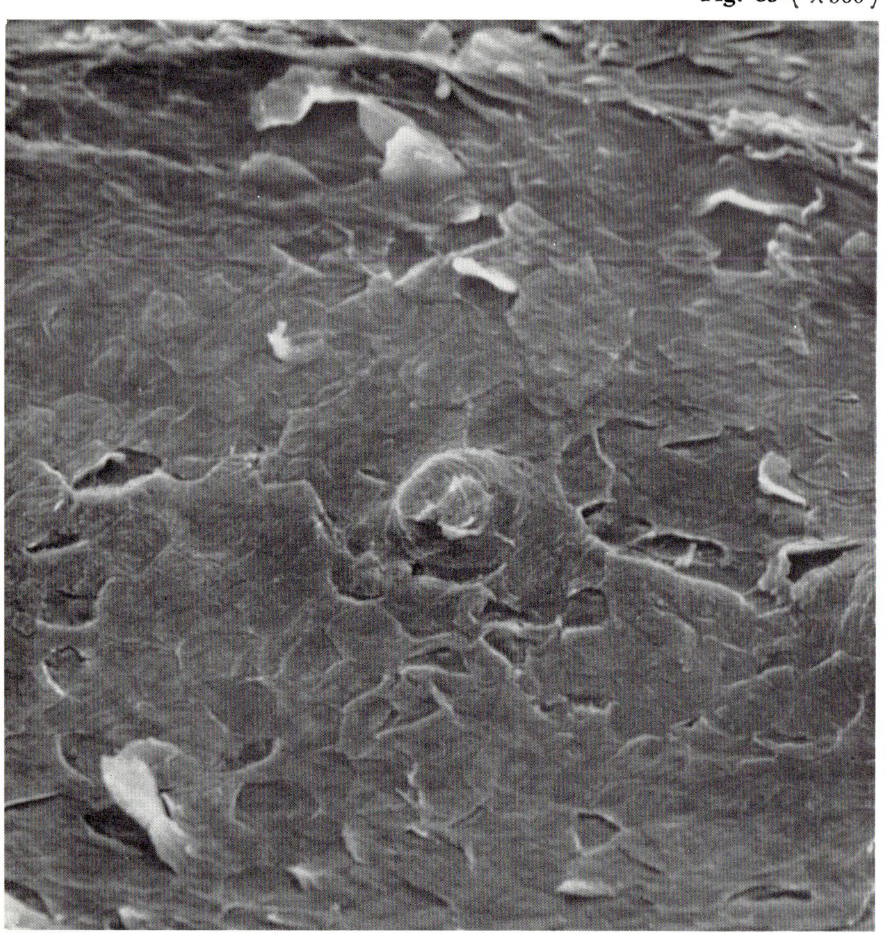

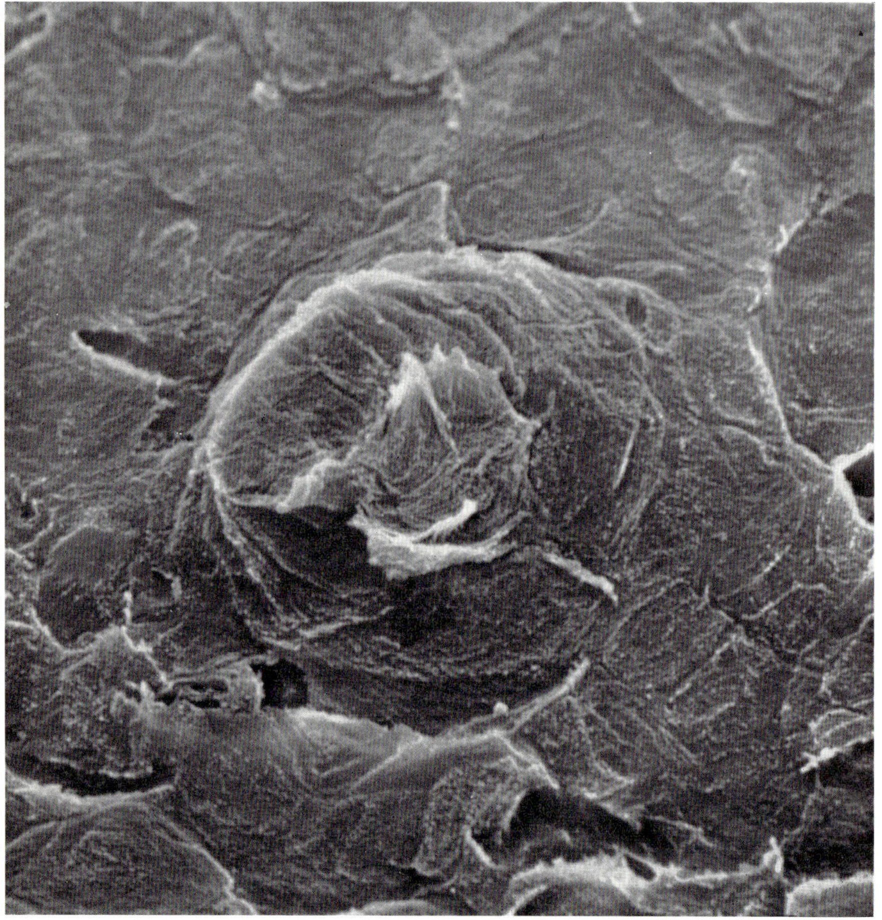

Fig. 86 (×1,200)

Figs. 85 and **87**, and the closer views in **Figs. 86** and **88** show the duct orifice in the center of the preceding pictures. The spiral course of the sweat gland duct is evident. The superficial cells of the skin show delicate furrows and pits.

Note

SUMP replicas of the finger belly of a 39-year-old man were coated with carbon and gold. EM: JSM-2

Fig. 87 (×360)

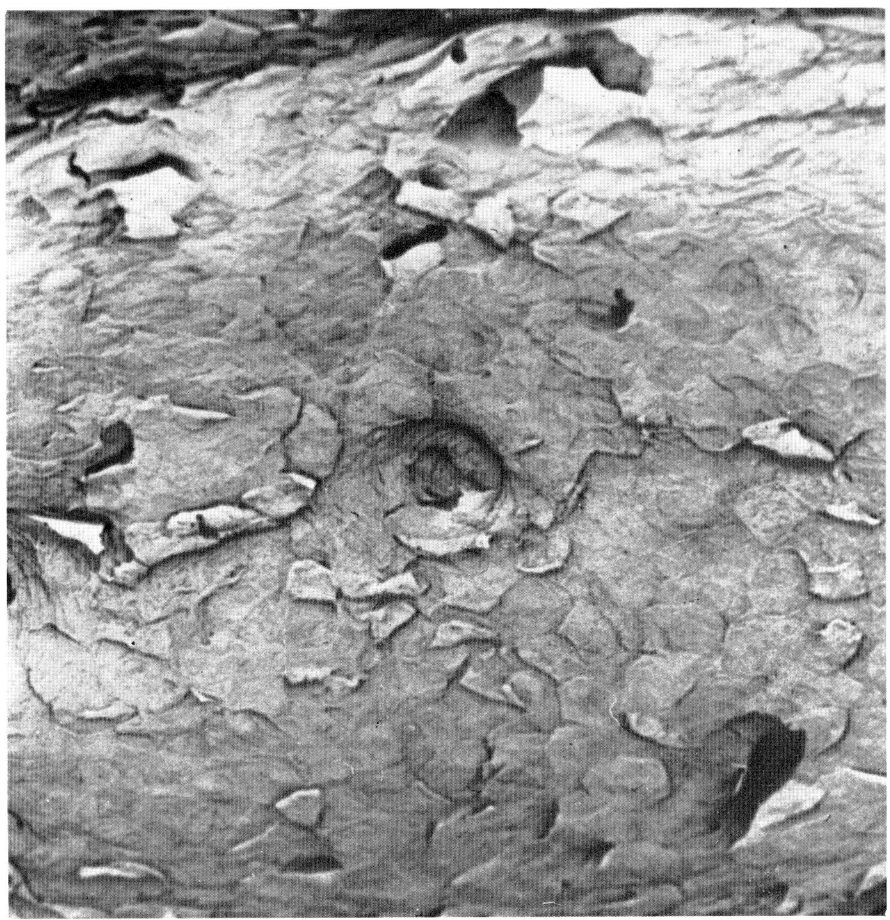

Fig. 88 (×1,200)

Reference

BERNSTEIN, E. O. and C. B. JONES: Skin replication procedure for the scanning electron microscope. Science 166: 252-253 (1969).

FUJITA, T., J. TOKUNAGA and H. INOUE: Scanning electron microscopy of the skin using celluloid impressions. Arch. histol. jap. 30: 321-326 (1969).

Figs. 83–88 by courtesy of the Arch. histol. jap.

21. Corti's Organ of the Inner Ear

The organ of Corti is a complex of thickened cells in the cochlea of the inner ear and is the effective organ of hearing. Lying on the basilar membrane which divides the spiral duct of the cochlea, it is comprised of supporting cells of various forms and of hair cells, the receptors of stimuli produced by sound. The hair cells are arranged in one inner row and three or, in a part of the cochlea, four outer rows. A hood-like structure called tectorial membrane covers the upper surface of these haired and supporting cells which is called reticular membrane (**Fig. 92**).

Fig. 89 shows the guinea-pig reticular membrane exposed by elimination of the tectorial membrane. In the lower left part of the picture the apical surface of the inner hair cells are seen. The zone "P" corresponds to the head plate of supporting cells called pillar cells. Above this zone there are three rows of outer hair cells. The apex of the first row cells is a uniquely modified triangle whereas that of the other two rows is octagonal. The hairs of each cell form a V or U which is especially obvious in the first row cells in this picture.

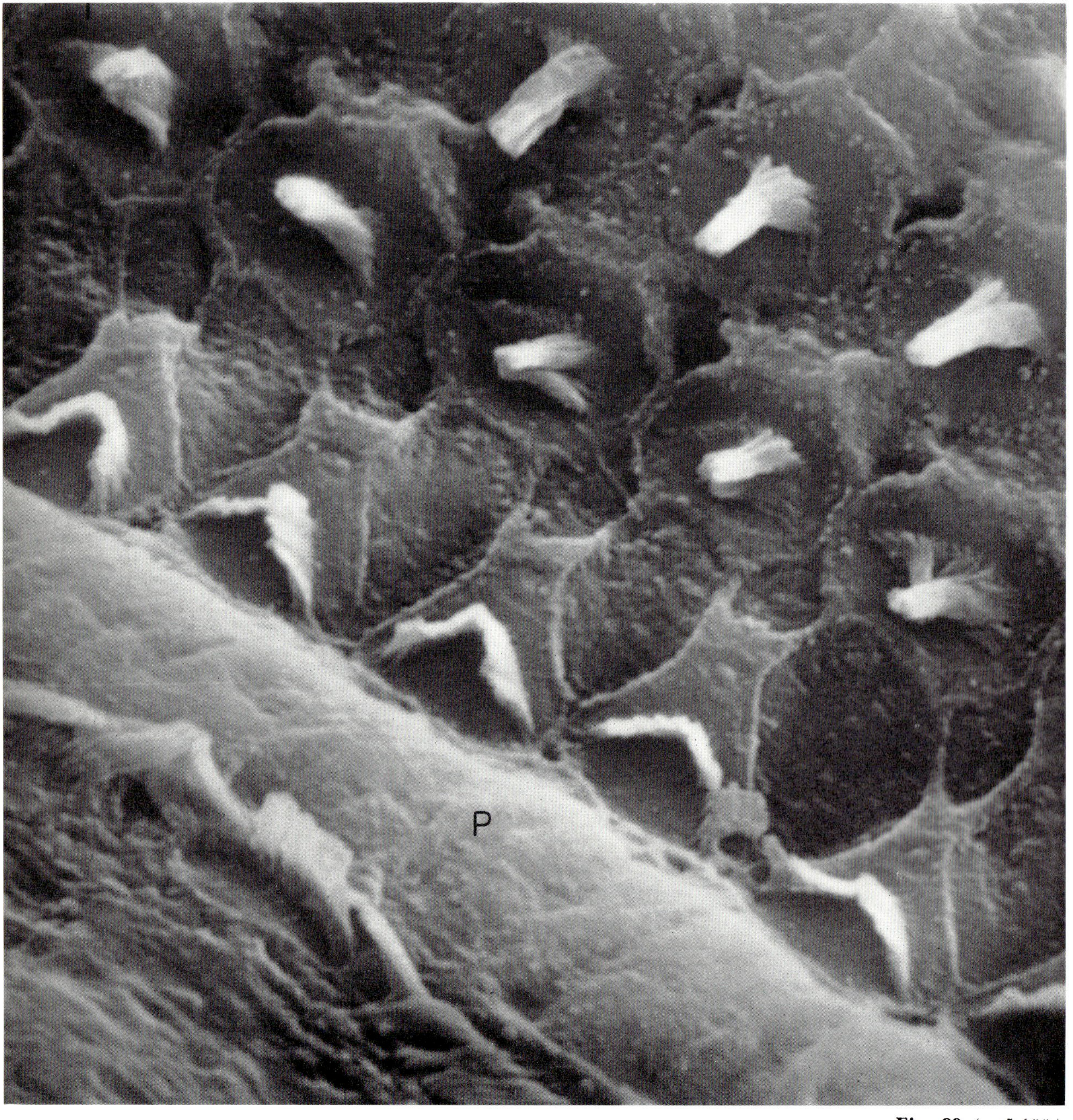

Fig. 89 (×5,100)

Fig. 90 is a closer view of the apex of an outer hair cell of the first row. The hairs, appearing like the teeth of a comb, are shorter in the center of the comb than at its ends. The small ball indicated by an arrow is the basal body of this cell.

Fig. 91 shows the apex of an inner hair cell at a high magnification. Three rows of hairs are labeled "1", "2" and "3". The hairs are long in the outermost row and rapidly become shorter towards the inner rows which are seen in the front in this picture.

The scanning electron micrographs of this section were provided by courtesy of Dr. NAOYA KOSAKA, Department of Otorhinolaryngology of Okayama University School of Medicine.

Note

Corti's organ excised from the cochlea of a guinea-pig was fixed in 10 per cent formalin, dehydrated in acetone, dried in air and coated with carbon and gold. EM: JSM-2

Reference

ENGSTRÖM, H., H. W. ADES and J. E. HAWKINS Jr.: Structure and functions of the sensory hairs of the inner ear. J. Acoust. Soc. Amer. 34: 1356-1363 (1962).

LIM, D. J.: Three dimensional observation of the inner ear with the scanning electron microscope. Acta otolaryngol., Suppl. 255: 1-38 (1970).

Fig. 90 (× 12,000)

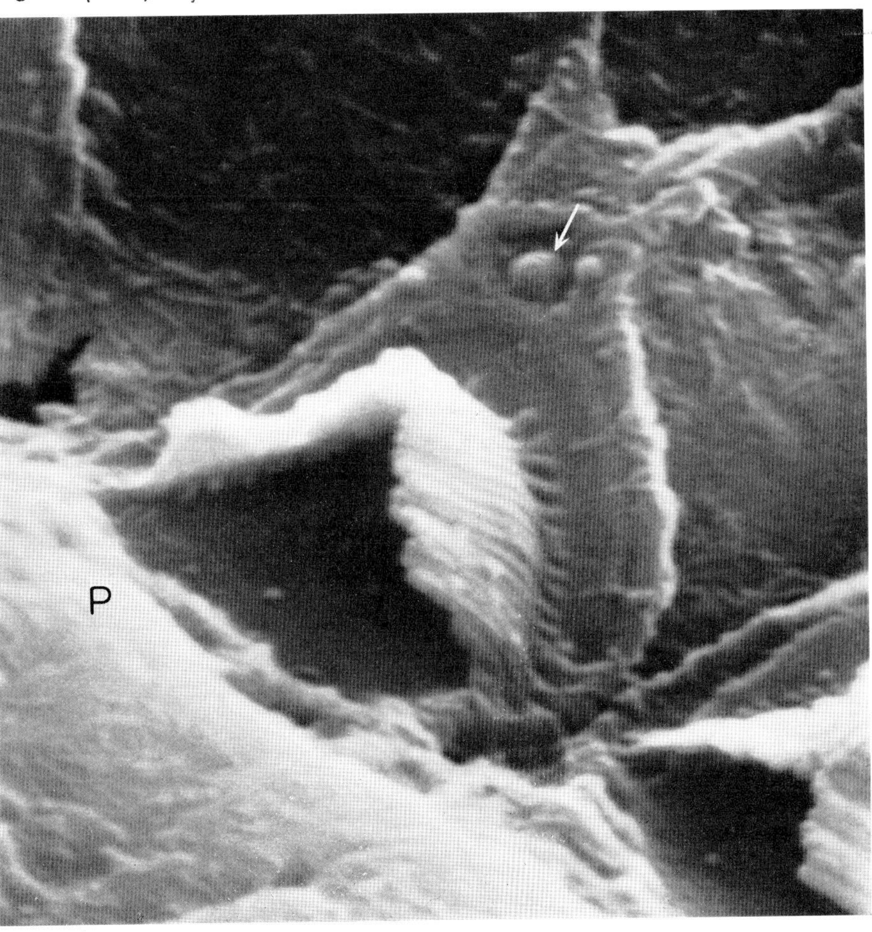

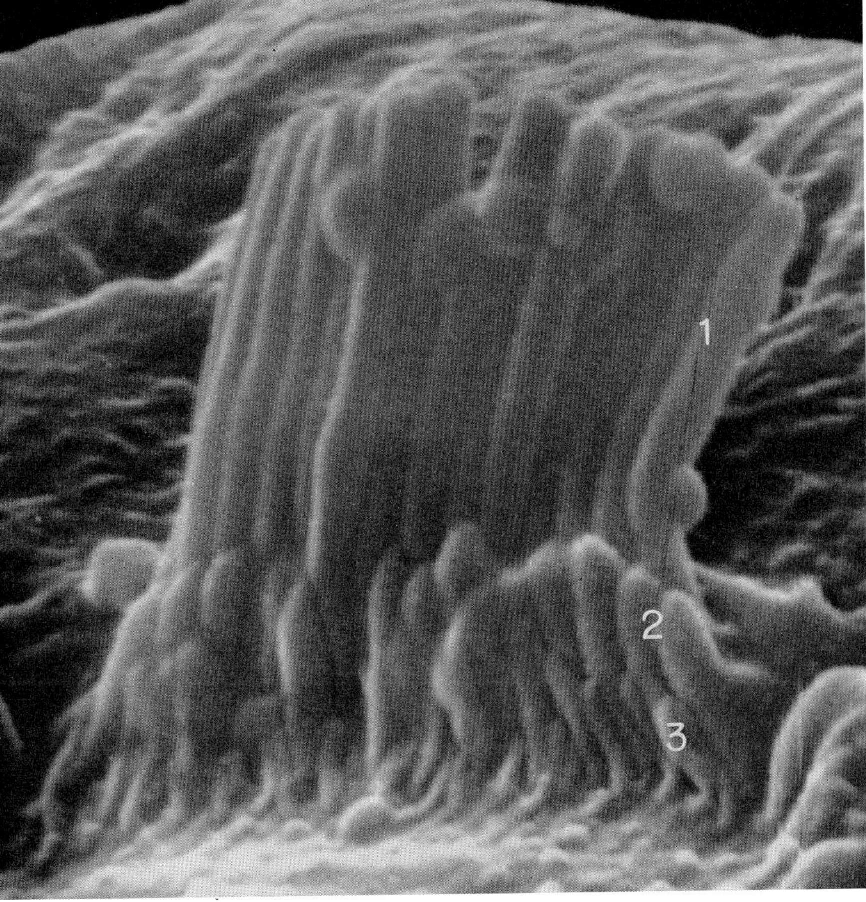

Fig. 91 (× 22,000)

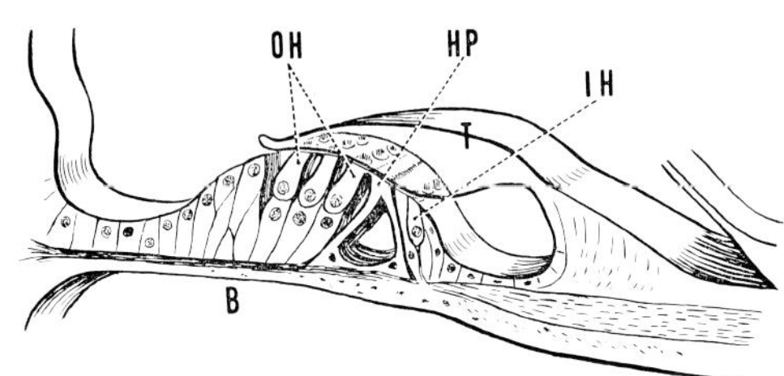

Fig. 92. A diagram showing the structure of Corti's organ.

B: basilar membrane. **T:** Tectorial membrane shown transparent so that the reticular membrane can be illustrated. **OH:** outer hair cells. **IH:** inner hair cell. **HP:** head plate of pillar cells.

22. Architecture of the Lens of the Eye

The lens of the eye is composed of a number of unusually long cells called "lens fibers." They run arched courses from the anterior to the posterior pole of the lens and are piled on each other like onion scales.

In mammals each fiber is a six-sided prism. From each of its crests, rod-like cytoplasmic processes protrude at regular intervals and are interdigitated with corresponding processes of adjacent fibers as shown in the diagram (**Fig. 95**).

Fig. 93 shows the lens fibers of the cattle. Fibers revealing their hexagonal cross-sections are densely piled up. In **Fig. 94** cytoplasmic processes on every crest of the fibers are evident.

The processes of lens fibers differ markedly in shape and arrangement from class to class of vertebrates, thus proposing a field of study highly interesting from a phylogenetic viewpoint.

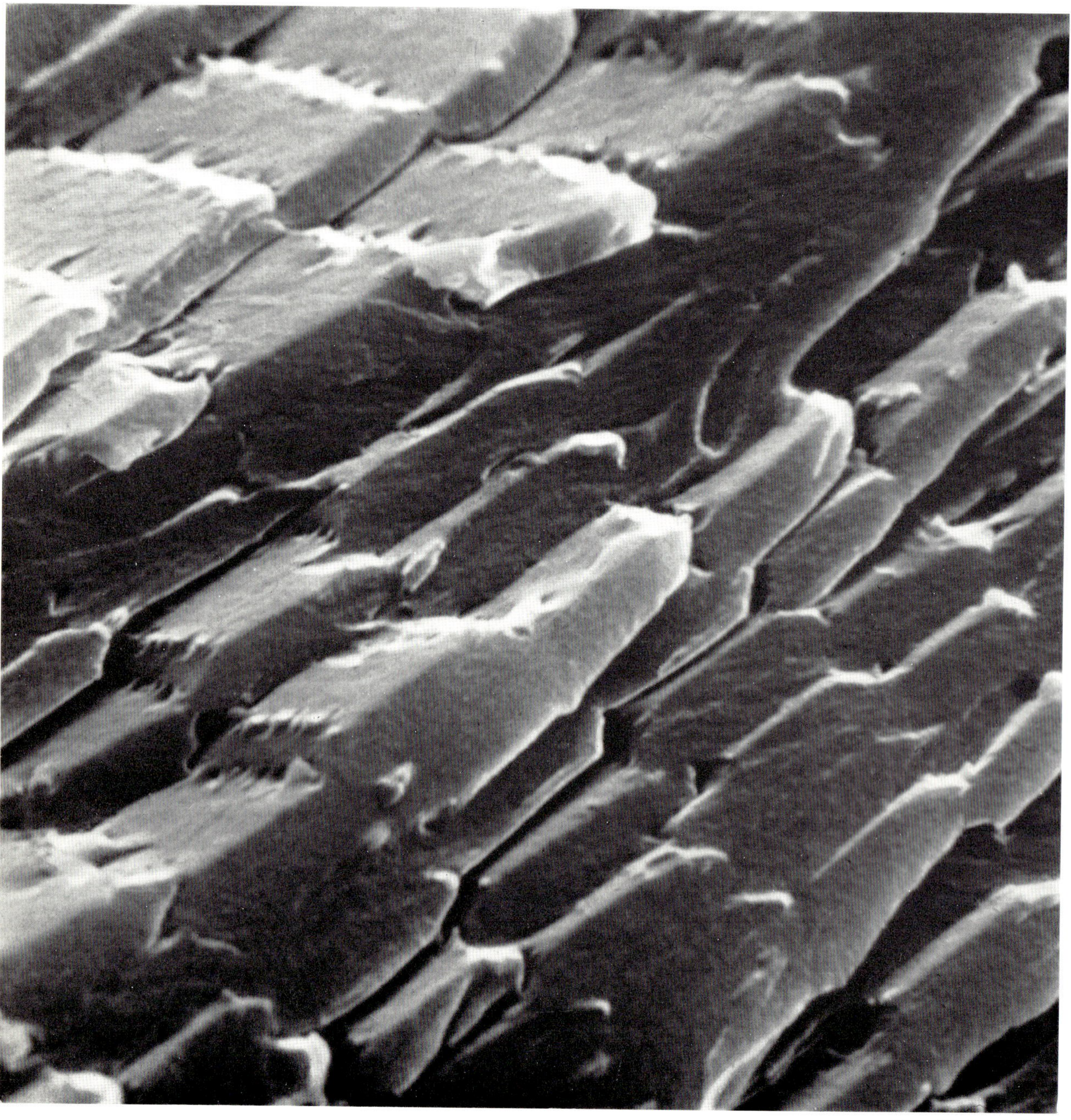

Fig. 93 (×5,000)

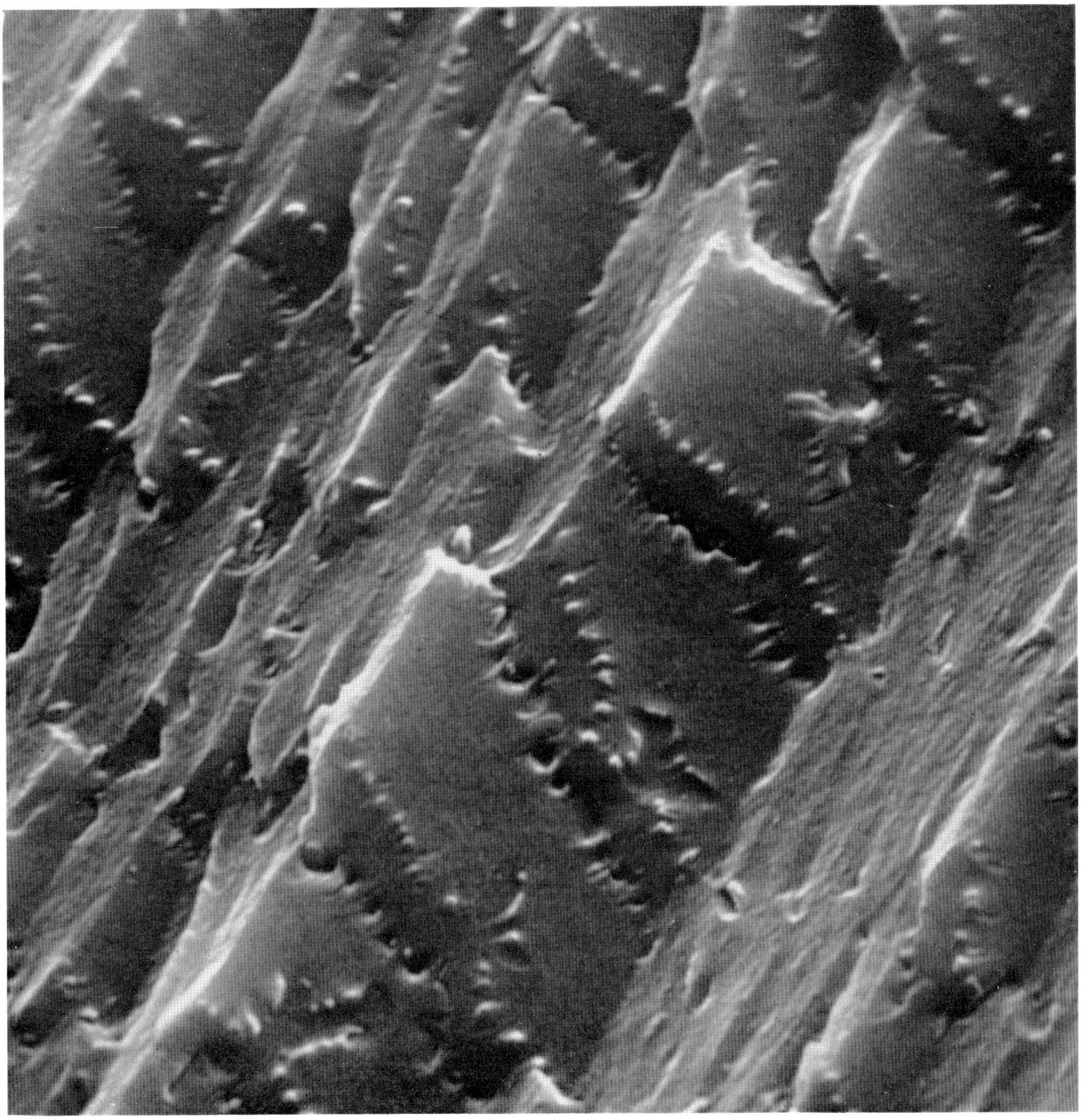

Fig. 94 (× 5,000)

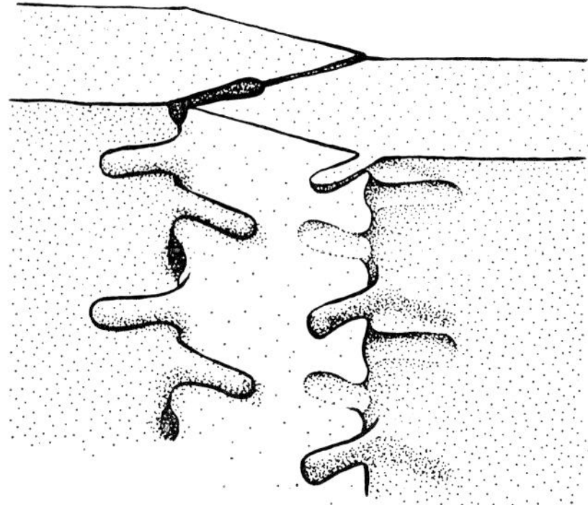

Fig. 95. A diagram showing the three-dimensional relations of lens fiber processes.

In mammals and other vertebrate classes in which the accommodation of the eye is based on the change in the thickness of the lens, the occurrence and interdigitation of the cytoplasmic processes of lens fibers are presumed to prevent excess aberration of individual fibers. This interpretation, however, seems not valid in the fish lenses which do not change their shape in accommodation and, nevertheless, show complicated interdigitation of large processes of lens fibers.

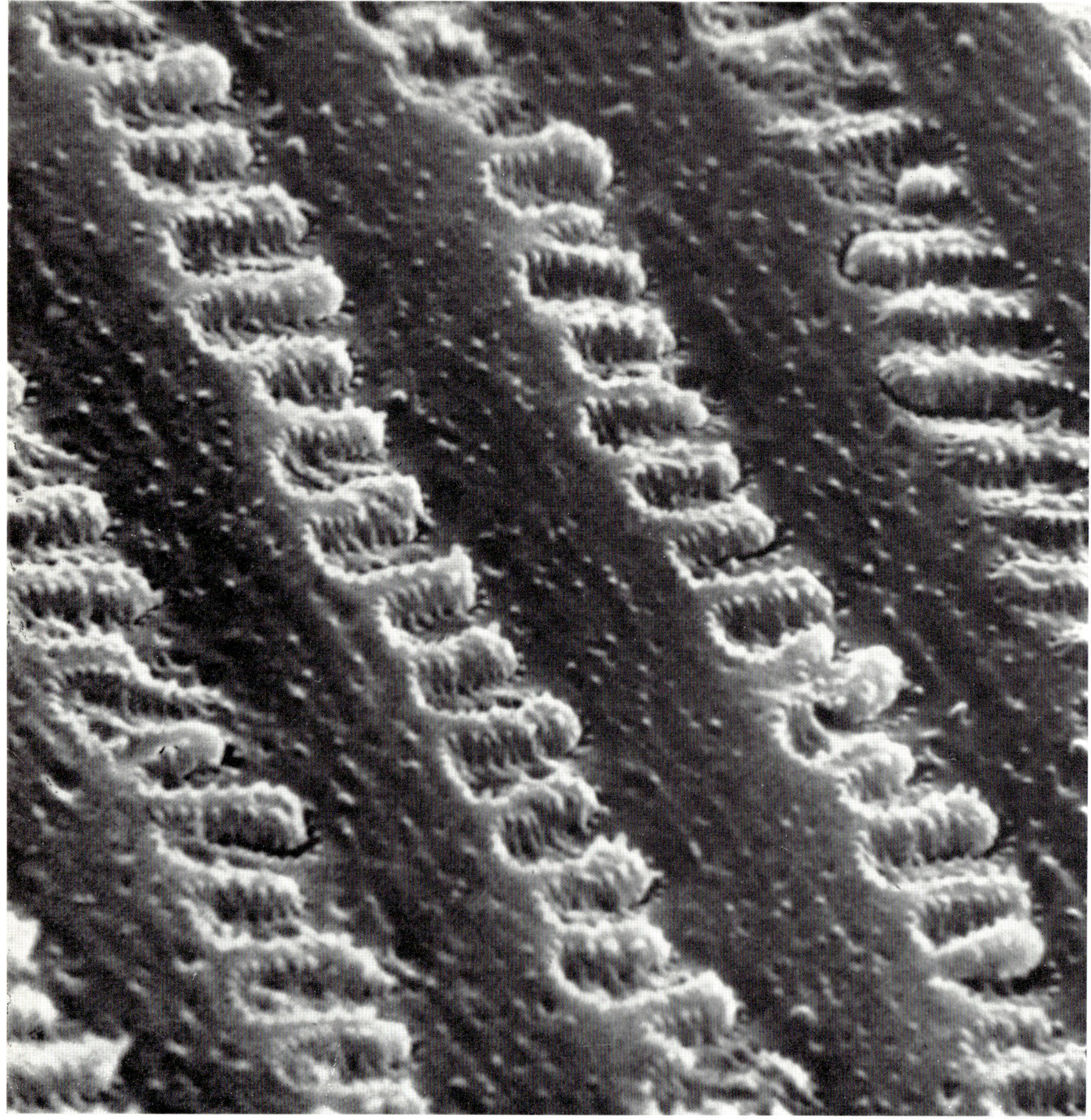

Fig. 96 (×5,000)

Teleosts, or bony fishes, have the most developed projections of lens fibers among vertebrates. **Fig. 96** is a scanning electron micrograph of the lens of the *Coryphaena hippurus*, one of the teleosts. **Fig. 97** shows a part of this picture at higher magnification. The edges of each fiber are alternately projected and indented, showing a deep interdigitation with adjacent fibers. The main processes are covered with small secondary processes.

Fig. 98 reveals, in a vertically cut portion, how the main processes of fibers are piled at an identical phase, layer on layer.

The figures and data of this section were provided by courtesy of Dr. Kéiichi Tanaka, Department of Anatomy of the Tottori University School of Medicine.

Fig. 97 (× 12,000)

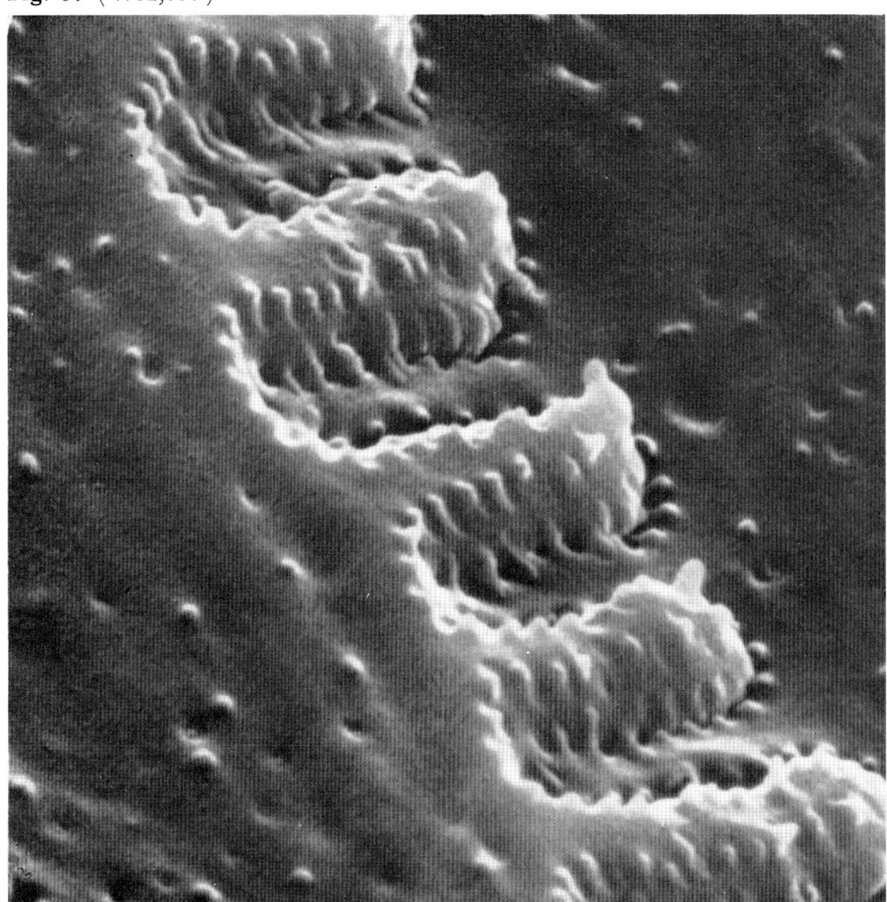

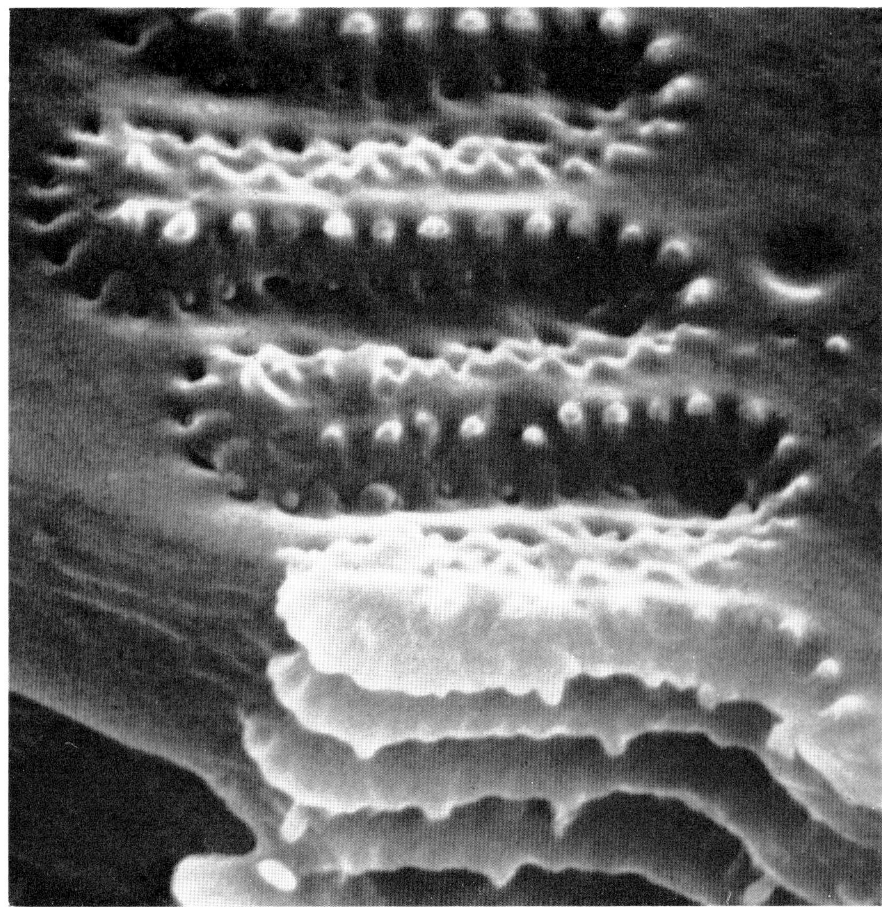

Fig. 98 (× 12,000)

Note

The lens was fixed in 10 per cent formalin and then kept in glycerin until it turned a deep amber in color. It was left in air more than one week and excess glycerin was removed. Superficial fibers were peeled off by forceps and the layers newly exposed were excised and coated with copper. EM: JSM-2

Reference

TANAKA, K.: Darstellung der Linsenfasern von Fischen anhand von Abdrücken und mittels des Raster-Elektronenmikroskops. Arch. histol. jap. 30: 233-246 (1969).

TANAKA, K. and A. IINO: Zur Frage der Verbindung der Linsenfasern im Rinderauge. Z. Zellforsch. 82: 604-612 (1969).

Figs. 96–98 by courtesy of the Arch. histol. jap.

FREE CELLS

23. Mast Cells and their Degranulation

Mast cells are dispersed in the connective tissue of the body and their cytoplasmic granules contain histamine, an important biogenic amine mediating the occurrence of allergic reactions.

Numerous mast cells occur floating in the peritoneal fluid of the rat, and can be used as an adequate material of scanning electron microscopy. **Fig. 99** shows some of

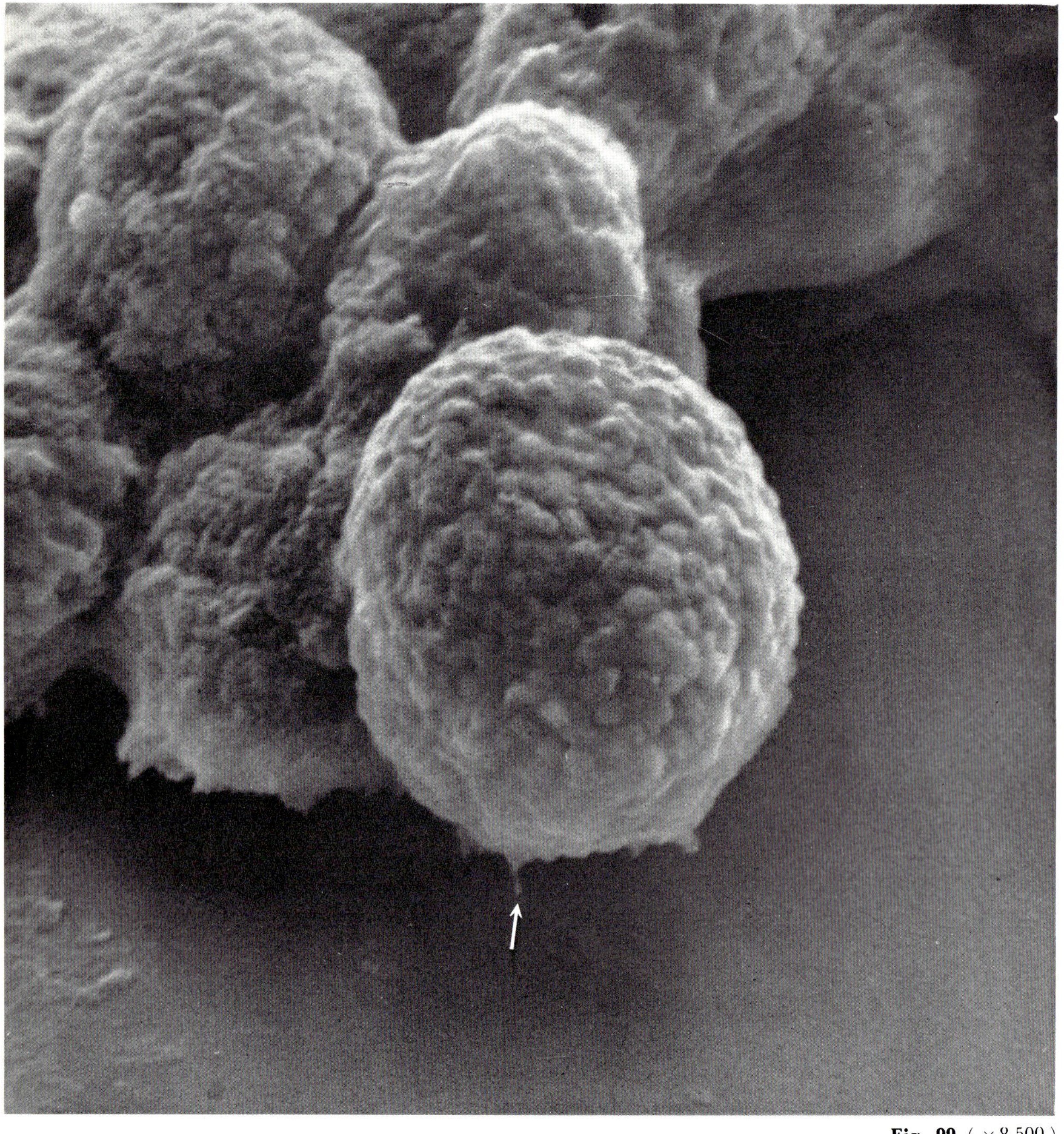

Fig. 99 (×8,500)

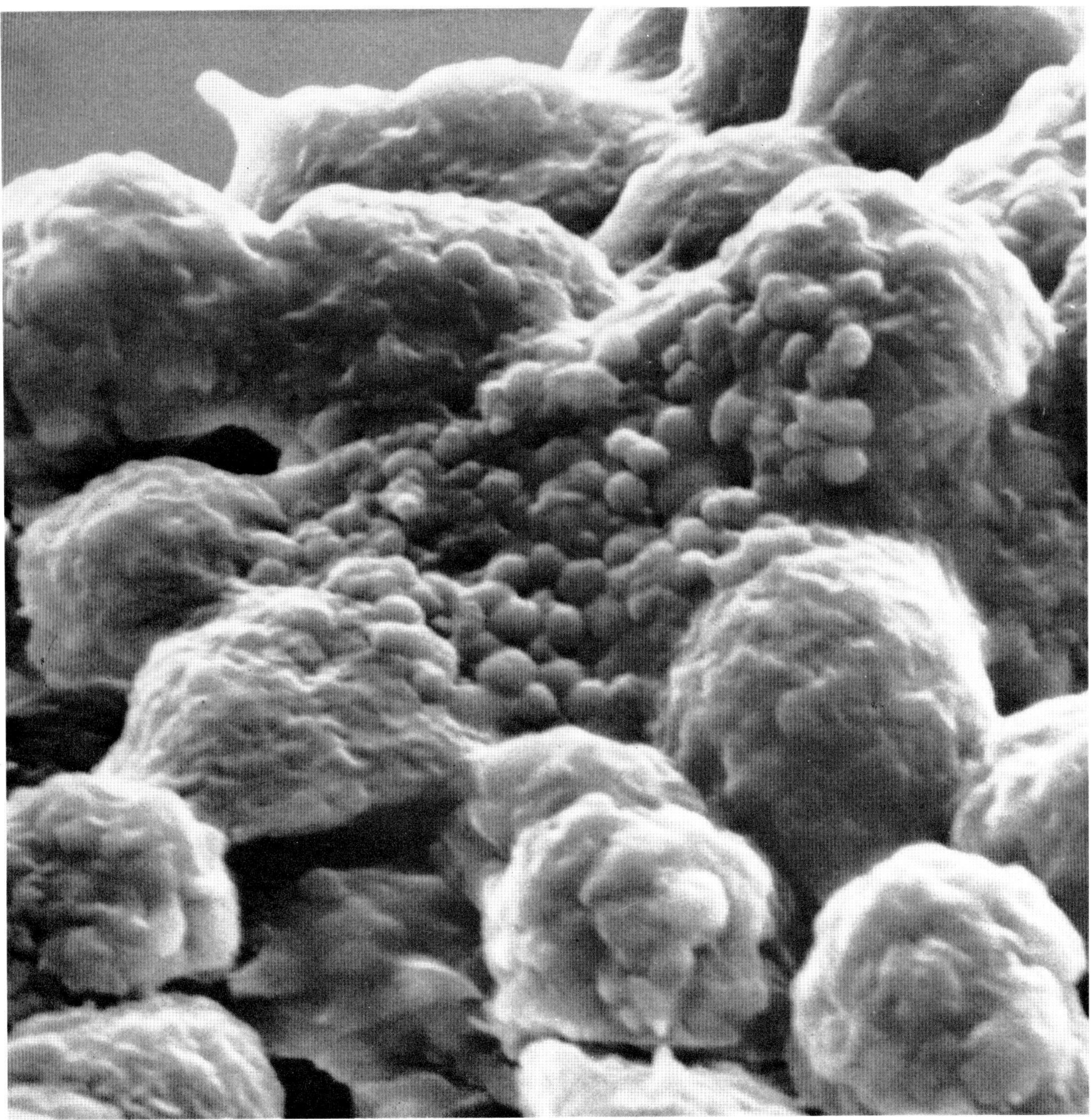

Fig. 100 (× 10,000)

those mast cells. The round granules densely filling the cytoplasm are suggested by the relief on the surface. Though mast cells possess microvilli, they probably lie adhered to the cell surface and only a few can be seen (arrow).

A group of substances called histamine liberators acts on the mast cells causing their degranulation and release of histamine. **Fig. 100** shows the changes in mast cells after administration of one of these substances, Compound 48/80. Denuded granules are scattered in the center of the picture. In some other cells granules appear protruded though still covered by a cytoplasmic sheet.

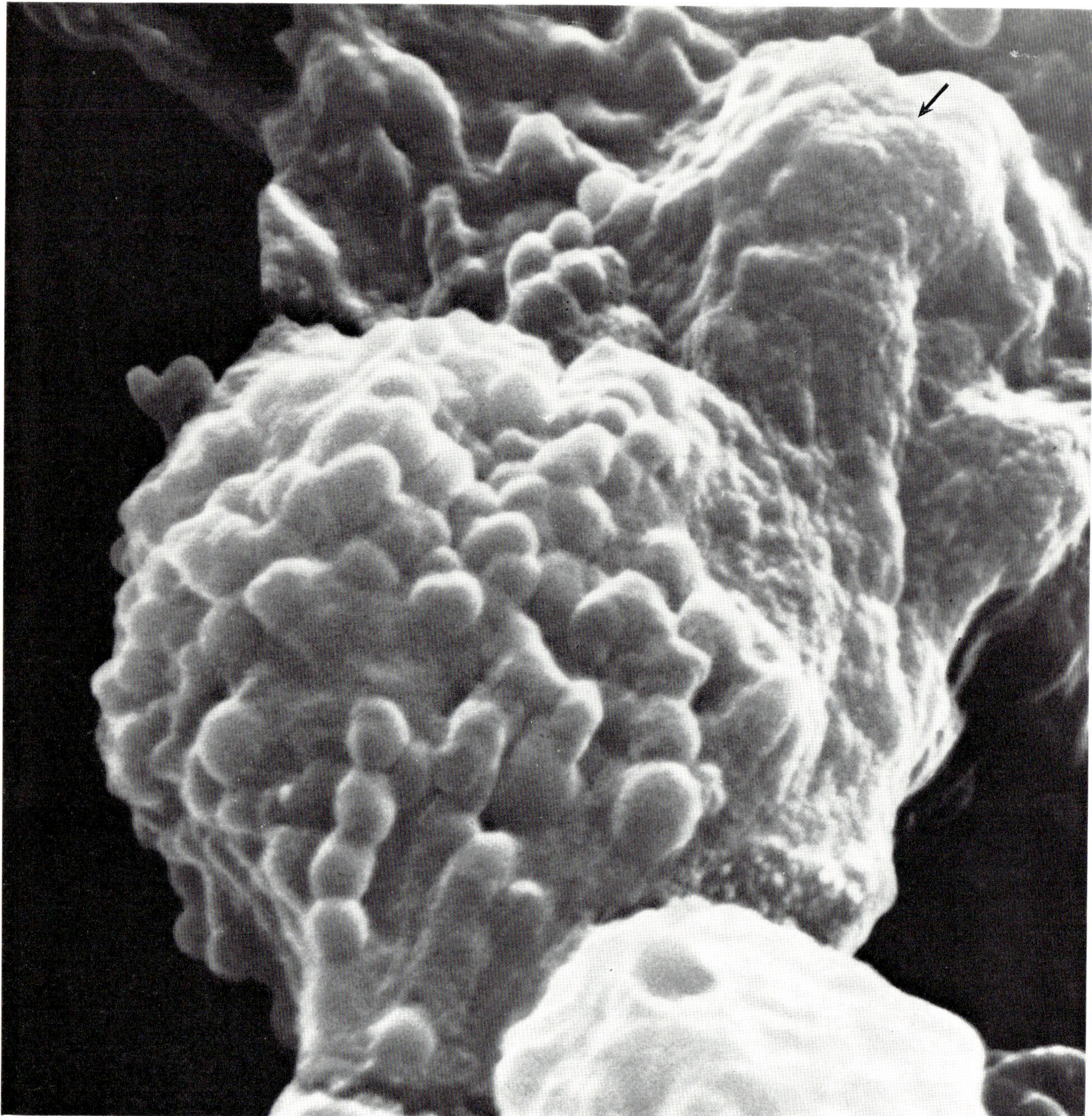

Fig. 101 (× 17,000)

Mast cell degranulation and histamine release occur also when antigen-antibody reaction takes place in the medium surrounding the cell. **Fig. 101** shows the change of rat mast cells after being mixed with anti-rat rabbit serum. In the cell shown in the center, granules are protruded because the covering cytoplasmic sheet has sunk due to the development of large vacuoles around the granules which is confirmed by transmission electron microscopy of sections made from the same material (YAMASAKI et al., 1970). Another cell with more advanced damage is seen in the upper part. The whole cell appears like a collapsed sac showing a swelling probably corresponding to the nucleus (arrow).

Figs. 102 and **103** show completely destroyed cells whose granules are partly denuded but seem still partly covered by a very thin cytoplasmic sheet. It is noticeable that the mast cell granules which are mainly composed of mucopolysaccharide seem to have a considerable consistency and that rugby-ball shaped granules are unexpectedly numerous.

Fig. 102 (× 12,000)

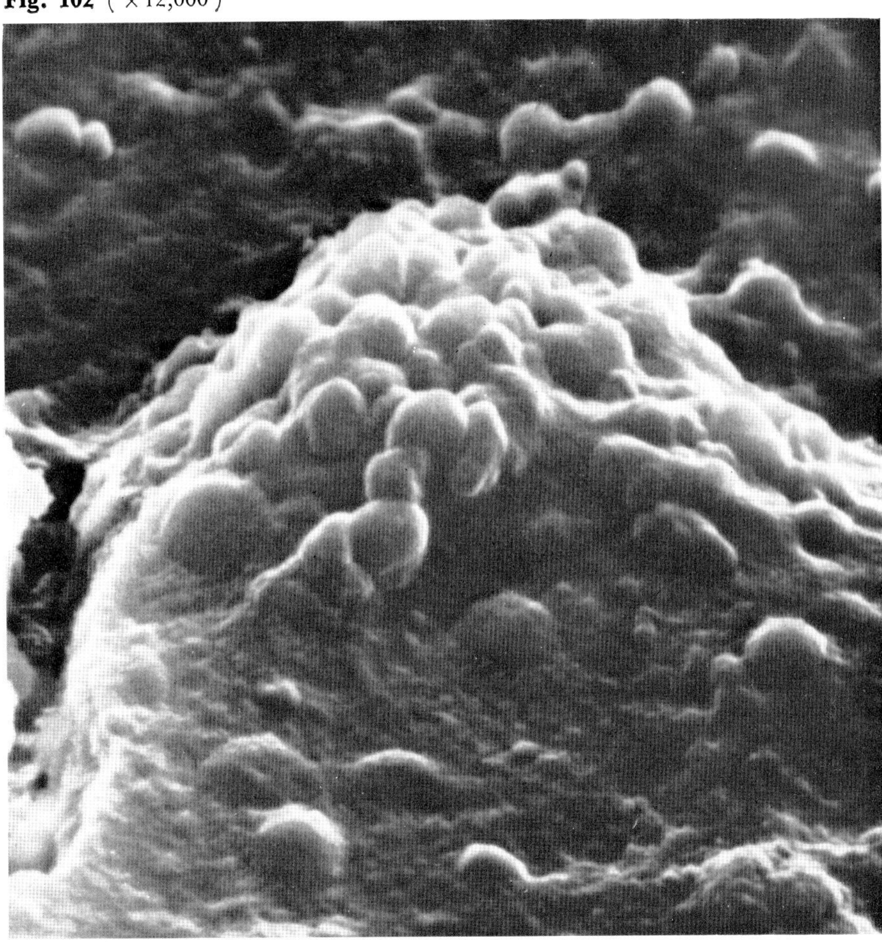

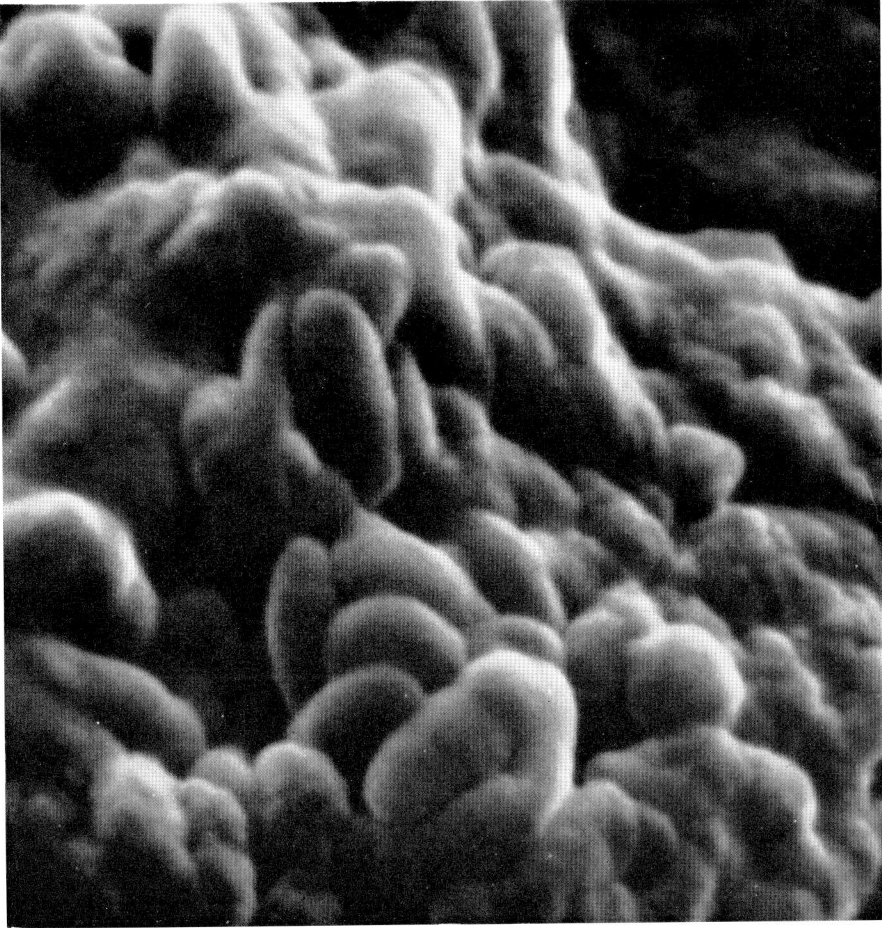

Fig. 103 (× 36,000)

Note

Physiological saline (10 ml) was injected into the peritoneal cavity of the rat. After massage of the abdomen the peritoneal cavity was opened and its fluid was taken with a pipet and gently centrifuged. The sediment was fixed in 2.5 per cent glutaraldehyde.

In the degranulation experiments the sediment was mixed either with 1 μg/ml compound 48/80 (Fig. 100) or with a 10 per cent v/v solution of anti-rat rabbit serum (Figs. 101–103) and incubated at 37°C for 15 minutes before fixation.

After being dehydrated in acetone the cells were dried on glass slides and coated with carbon and gold. EM: JSM-2

Reference

YAMASAKI, H., T. FUJITA, Y. OHARA and S. KOMOTO: Electron microscope studies on the release of histamine from rat peritoneal mast cells. Arch. histol. jap. 31: 393-408 (1970).

24. Human Blood Cells

The red blood cells are discs with concave surfaces. **Fig. 104** shows the appearance of normal human red cells. Their flexible forms and considerable variety in size may be impressive.

Noteworthy is the frequent occurrence of wart-like projections which seem to favor the peripheral portion of the cell. The cell in the center of **Fig. 105** show one large wart and another inconspicuous one, whereas the cell in the upper right has three small ones. These warts likely correspond to the vesicular appendages of red cells recognized by transmission electron microscopy (Miyoshi and Matsukura, 1969).

Fig. 106 shows a white blood cell found among red cells. The study with the scanning electron microscope has not yet advanced so far that different types of leucocytes have been identified by their surface structure.

Fig. 104 (×8,000)

Fgi. 105 (×15,000)

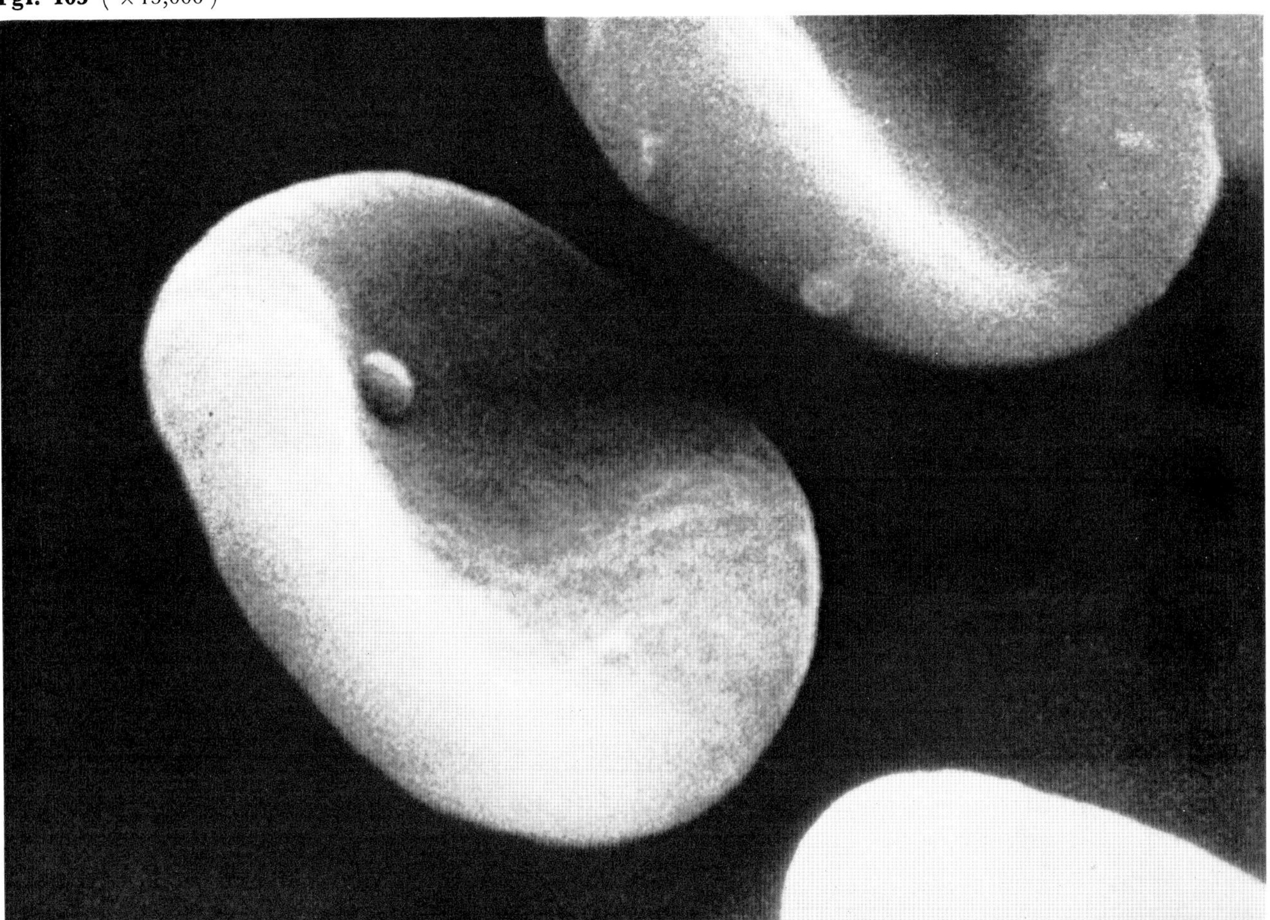

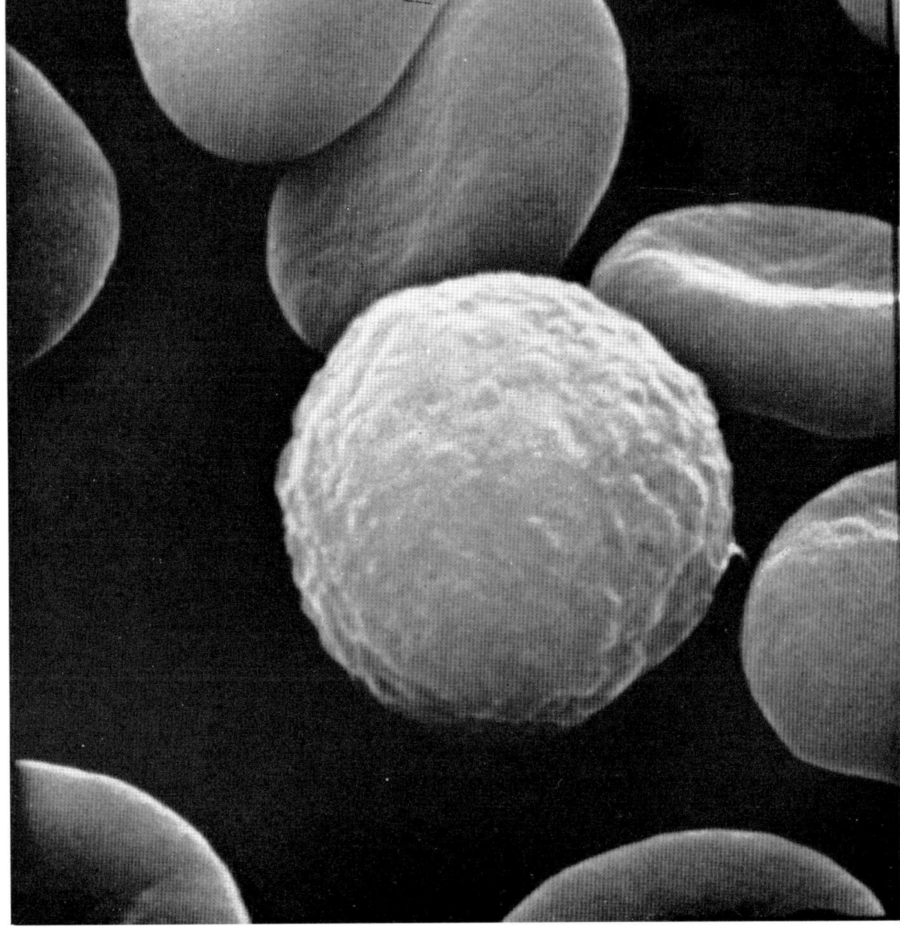

Fig. 106 (×8,600)

Note

Blood of a 39-year-old healthy man was fixed in 1 per cent glutaraldehyde (0.1M phosphate), according to the method shown in p. 2, dehydrated in acetone, dried on glass slides and coated with carbon and gold. EM: HSM-2 (Figs. 104 and 105) and JSM-2 (Fig. 106)

Reference

CLARKE, J.A. and A.J. SALSBURY: Surface ultrastructure of human blood cells. Nature 215: 402-404 (1967).

MIYOSHI, M. and Y. MATSUKURA: On the occurrence and fine structure of an appendage in human and murine erythrocytes. Arch. histol. jap. 31: 55-58 (1969).

SALSBURY, A. J. and J. A. CLARKE: New method for detecting changes in the surface appearance of human red blood cells. J. clin. Pathol. 20: 603-610 (1967).

TOKUNAGA, J., T. FUJITA and A. HATTORI: Scanning electron microscopy of normal and pathological human erythrocytes. Arch. histol. jap. 31: 21-35 (1969).

Fig. 105 by courtesy of the Arch. histol. jap.

25. Red Blood Cells and Osmotic Pressure

Red blood cells are known to be highly sensitive to the osmotic pressure of their medium. In a hypertonic circumstance they become flattened or crenate, whereas in a hypotonic one they swell into rounded forms.

Fig. 107 shows red blood cells treated with 3 per cent sodium chloride prior to fixation in 1 per cent glutaraldehyde. The red cells are flattened and crumpled and look like potato chips. A corresponding change in the shape of red cells occurs when blood is fixed in 2.5 per cent or 5 per cent glutaraldehyde.

Fig. 107 (×5,100)

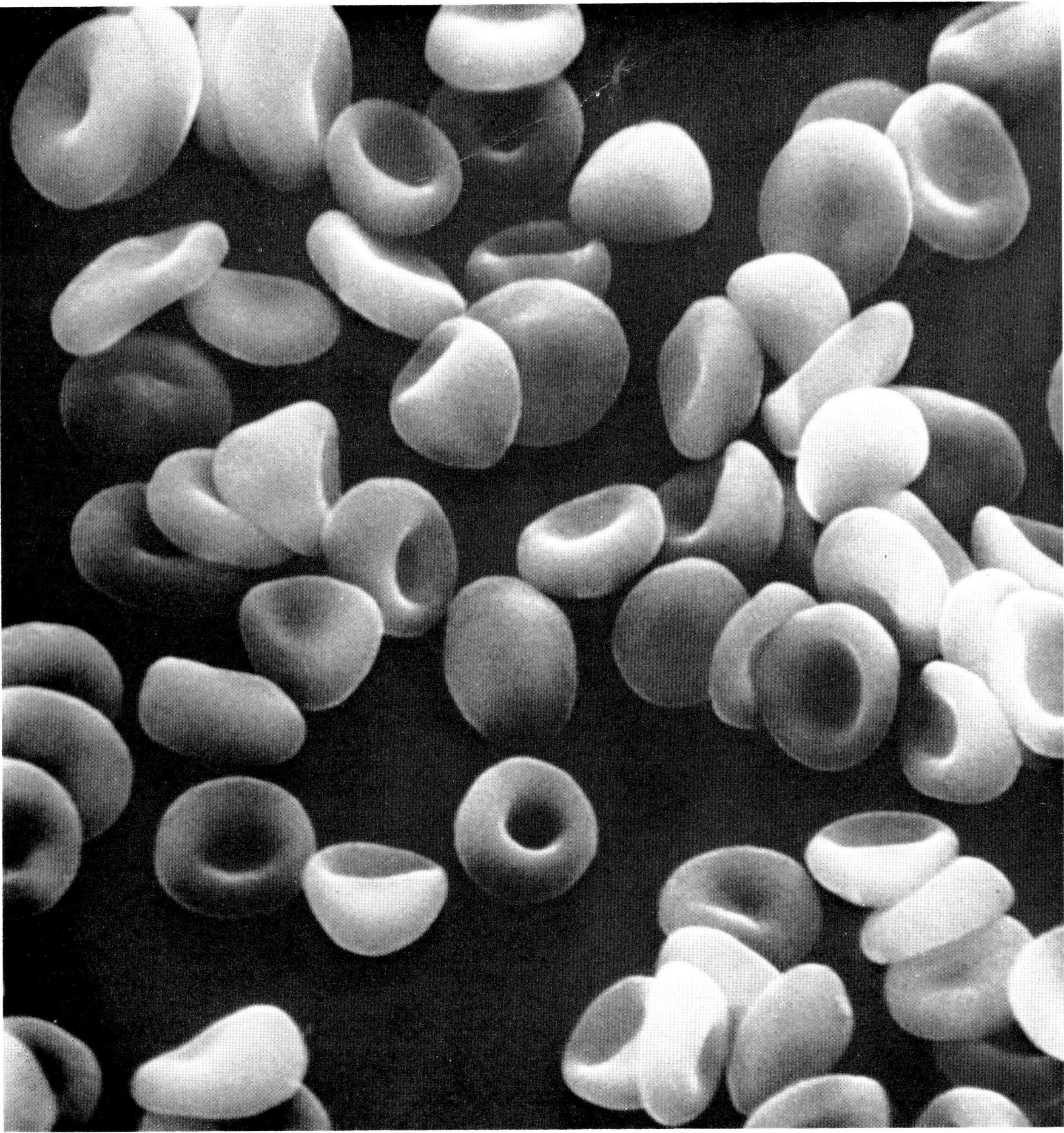

Fig. 108 (×5,100)

Fig. 108 shows swollen red blood cells after fixation in 0.5 per cent glutaraldehyde. There are seen rounded cells of bowl-like form in which only one of the concavities remains and the other side is strongly convex. Doughnut-shaped cells with thickened peripheral portion are also frequent. It is obvious that each cell has an individual reaction in response to the same osmotic pressure, and the change in shape may vary greatly from cell to cell.

Note

Fresh blood from a 39-year-old man was dropped in a 3 per cent NaCl solution, left for one hour and fixed in 1 per cent glutaraldehyde (0.1 M phosphate buffer). Fresh blood was fixed in 0.5 per cent glutaraldehyde (0.1 M phosphate buffer). The fixed materials were dehydrated in acetone, dried on glass slides and coated with carbon and gold. EM: JSM-2

Reference See p. 75.

26. Spherocytosis

Spherocytosis is an abnormality of red blood cells showing unusual variability in the shape and size of the cells, including spheroid forms.

The present patient is a 4-year-old boy with anemia and slight jaundice. The hereditary nature of spherocytosis is clear in this case as his father and brother have been diagnosed as having the same abnormality of blood.

Fig. 109 shows the red blood cells of this patient which are characteristically irregular in form. Although discoid shapes predominate, unusual small types are scattered throughout. Thickened cells, spheroid and crenate cells are also seen.

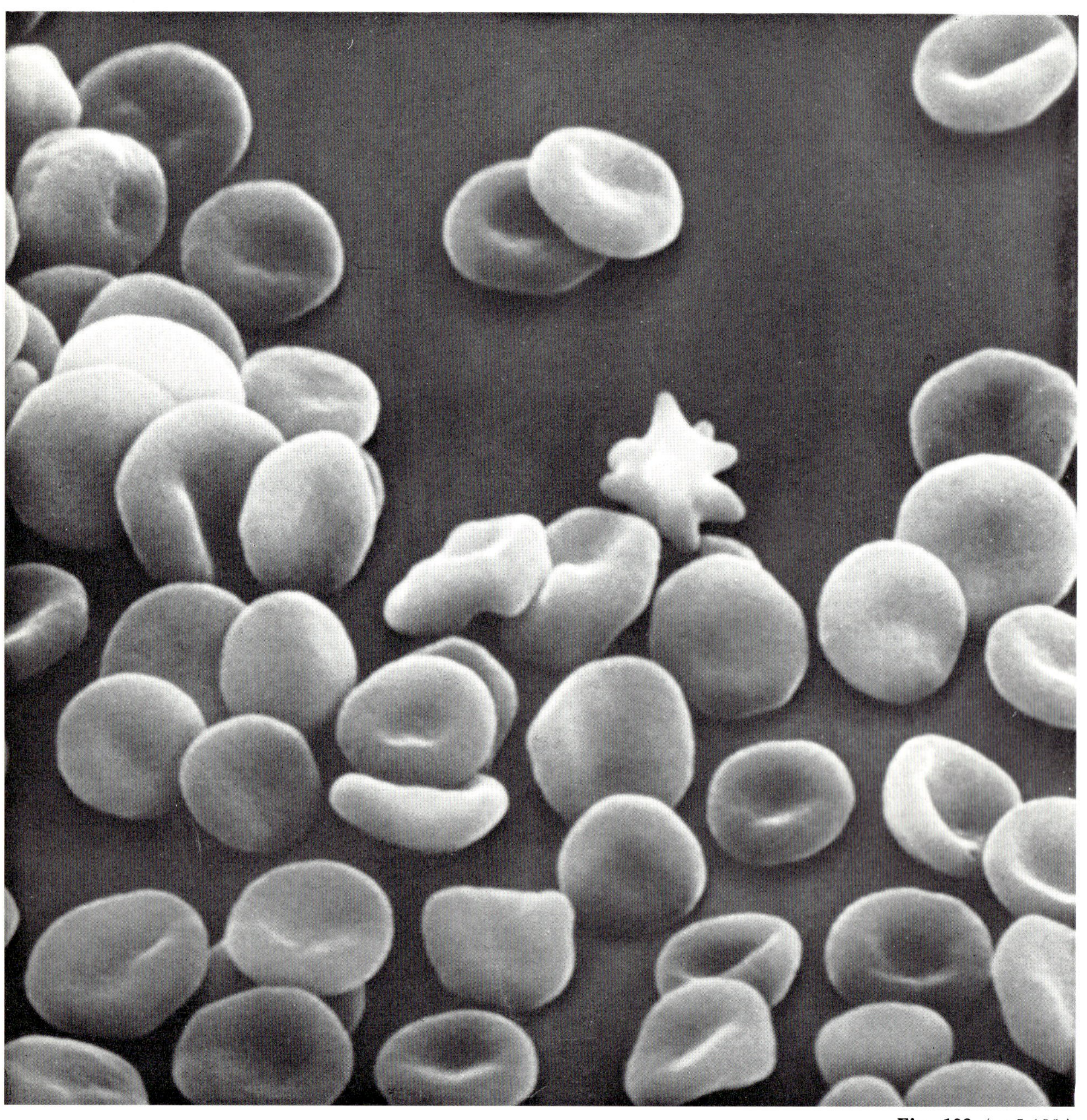

Fig. 109 (×5,100)

Fig. **110** (× 12,000)

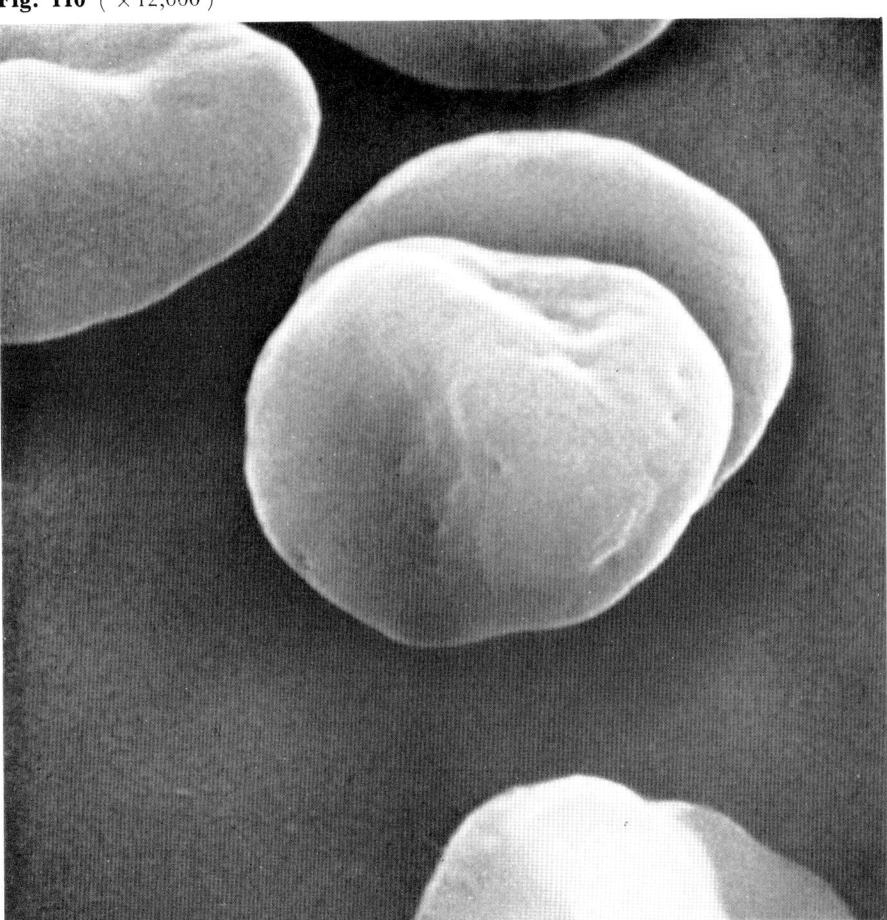

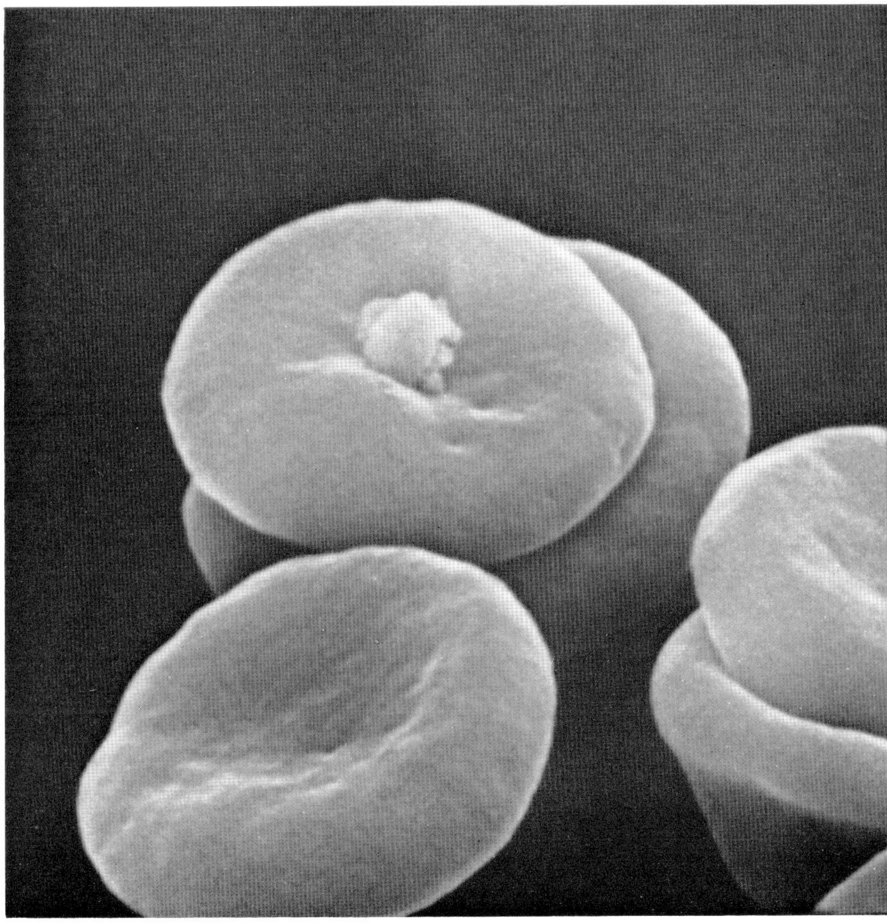

Fig. **111** (× 12,000)

At higher magnification (**Figs. 110** and **111**) small but apparently deep pits are recognized on the uneven surface of the red cells. SALSBURY and CLARKE (1967) who observed red blood cells of spherocytosis under the scanning electron microscope also recognized this peculiar pitting.

Note

Blood of the patient was fixed in 1 per cent glutaraldehyde (0.1 M phosphate buffer), dehydrated in acetone, dried on glass slides and coated with carbon and gold. EM: JSM-2

Reference

SALSBURY, A.J. and J.A. CLARKE: New method for detecting changes in the surface appearance of human red blood cells. J. clin. Pathol. 20: 603-610 (1967).

TOKUNAGA, J., T. FUJITA and A. HATTORI: Scanning electron microscopy of normal and pathological human erythrocytes. Arch. histol. jap. 31: 21-35 (1969).

27. Elliptocytosis

Elliptocytosis is an abnormality of blood characterized by the occurrence of elongated red blood cells. The enlarged spleen of the present patient, a 30-year-old male, was removed nine months prior to the present examination. Anemia and the other hemolytic signs he had suffered previously disappeared after the splenectomy.

As seen in **Fig. 112** the red blood cells of this patient deviate greatly from the normal shape and size. Elongated cells are mostly represented by elliptic plates.

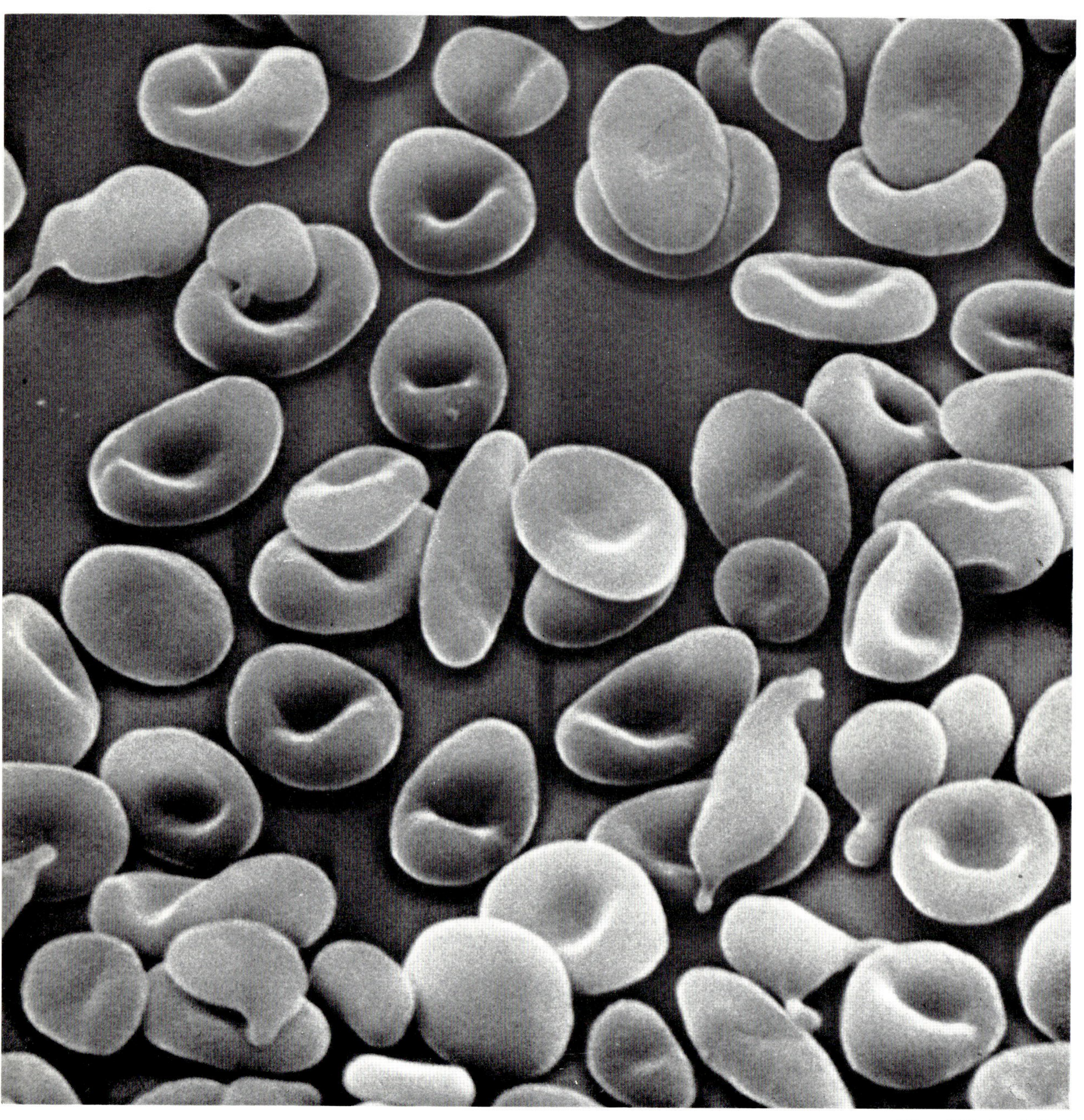

Fig. 112 (×5,100)

Fig. **113** (×6,000)

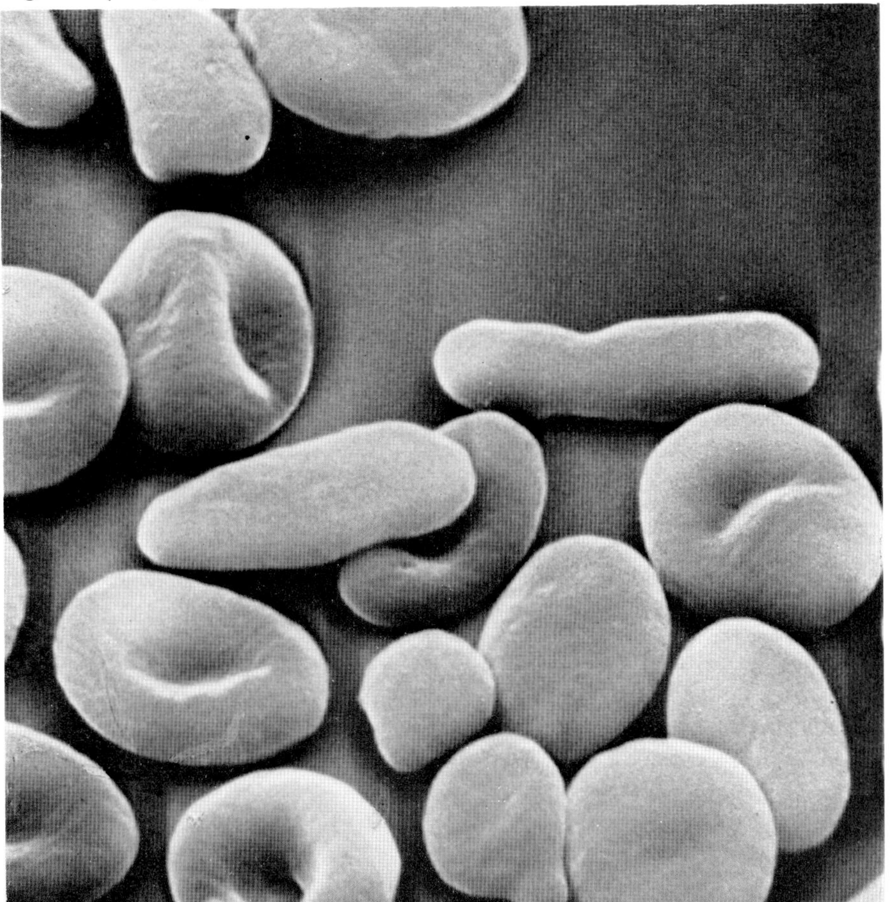

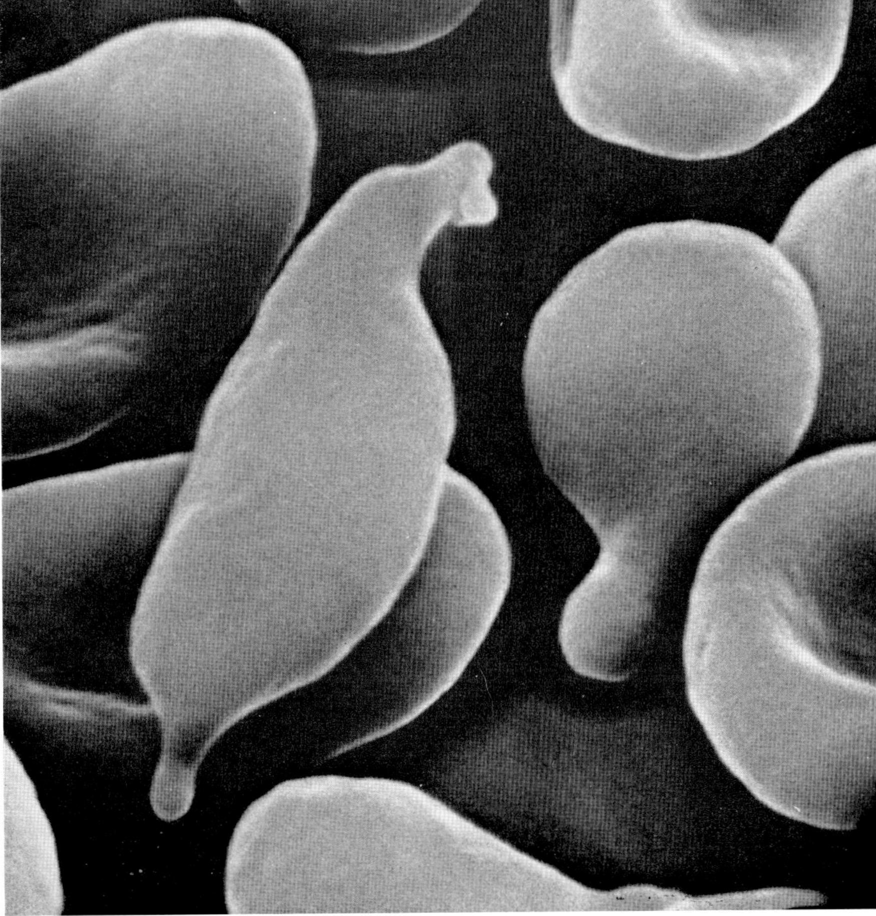

Fig. 114 (×12,000)

Fig. 113 shows conspicuously long rods. Though not every case of elliptocytosis is of hereditary nature, the occurrence of rod-shaped cells is said to be characteristic of the hereditary type.

Fig. 114 shows cells of irregular shapes showing thin and rounded processes. This unique type of cell is known to be common in typical elliptocytosis and is called a tear-drop cell.

Note and **Reference** See p. 79.

Figs. 107-114 by courtesy of the Arch. histol. jap.

28. Blood Smears

Clinical hematology is based on the microscopic examination of blood smears. Scanning electron microscopy impressively demonstrates that in blood smears the cells, especially leucocytes, are completely flattened like little pancakes and that all of them are half buried in the smooth layer of plasma which evenly covers the glass slides. The clear images of the nuclear lobes and granules of leucocytes which a blood smear affords is based on this artificial flattening of the cells.

Figs. 115, 116 and **117** respectively show an eosinophile, neutrophile and lymphocyte together with red blood cells. The nucleus which is lobulated in the former two and simply round in the last cell has sunk as compared with the cytoplasm.

A blood smear once stained by a hematological method can be coated with metals for scanning electron microscopy. A smear once examined in the scanning microscope may be observed again by light microscopy (McDonald, 1969).

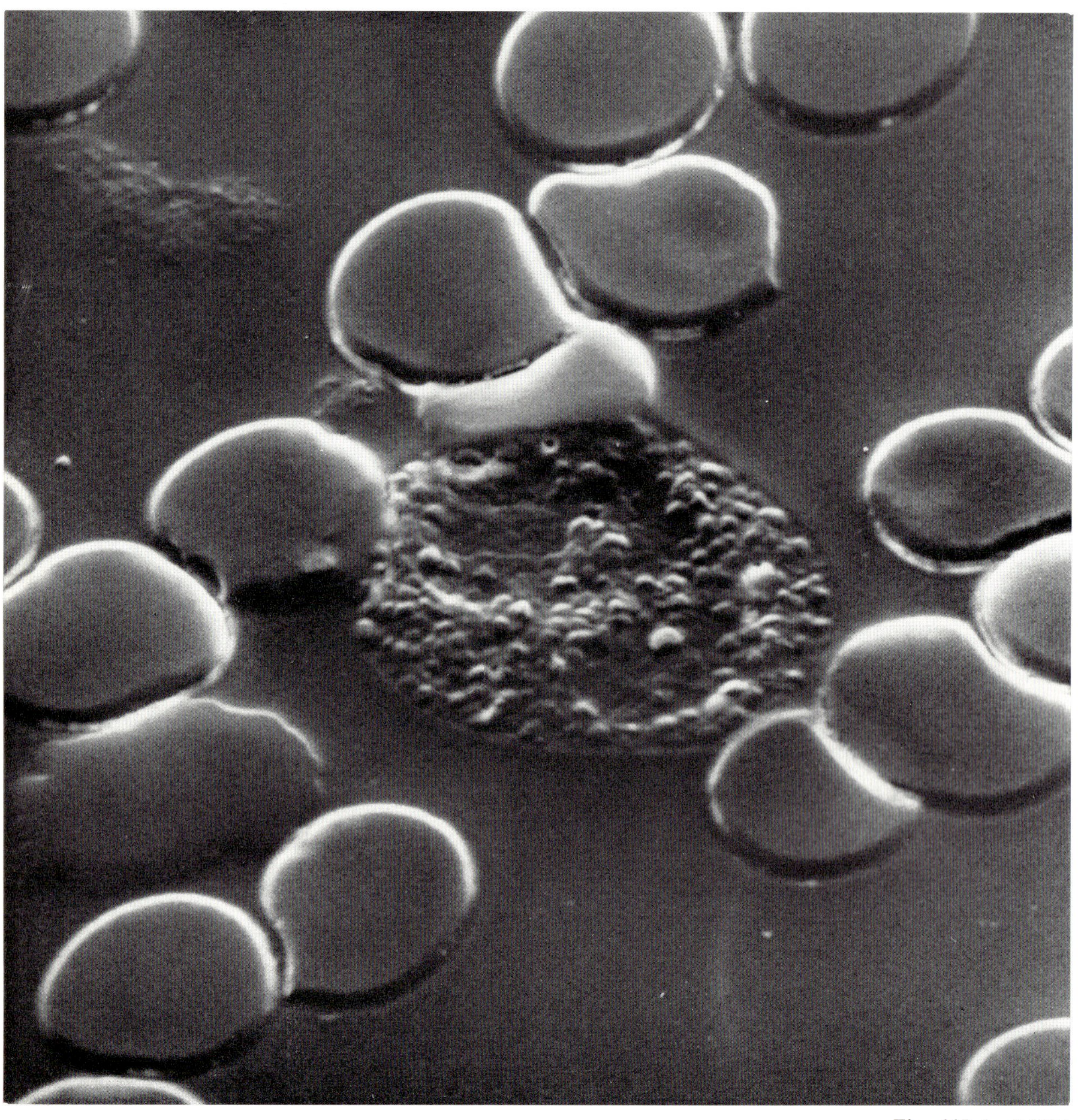

Fig. 115 (×5,100)

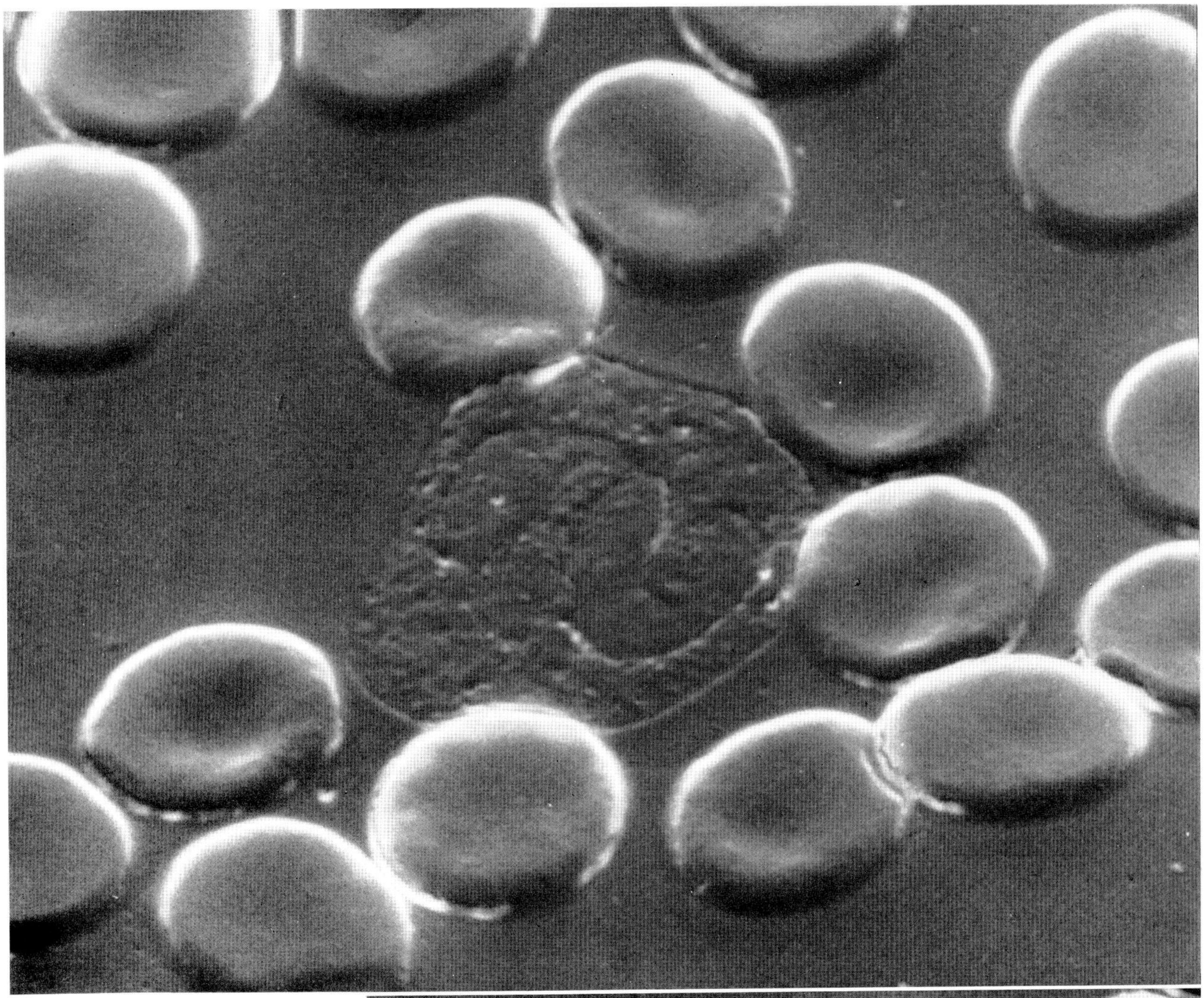

Fig. 116 (×5,100)

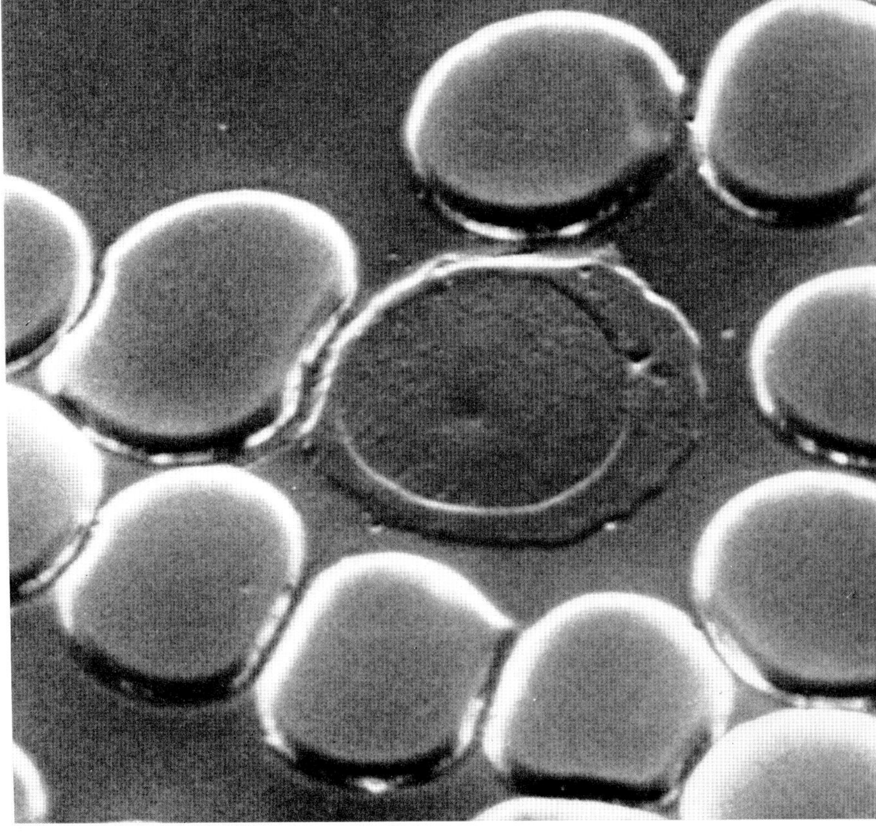

Fig. 117 (×5,100)

Thus the figures of one and the same cell can be observed alternately under both microscopes.

Note

Blood smears of a patient with chronic myelogenous leukemia were coated with carbon and gold. EM: JSM-2

Specimen preparation by Dr. AKIRA HATTORI, Department of Internal Medicine, Niigata University Medical School.

Reference

McDONALD, L. W.: Correlation of scanning electron microscope and light microscope images of individual cells in human blood and blood clots. Exp. molec. Pathol. 10: 186-198 (1969).

29. Red Blood Cells of Animals

Although the red blood cells in man and mammals are so specialized a type of cell that they have lost their nuclei during their maturation in the bone marrow, those in the lower classes of vertebrates are nucleated forms as are the general cells of the body.

Fig. 118 is a scanning electron micrograph of the red cells in the fowl. The cells are elliptic in shape as is generally the case in lower vertebrates.

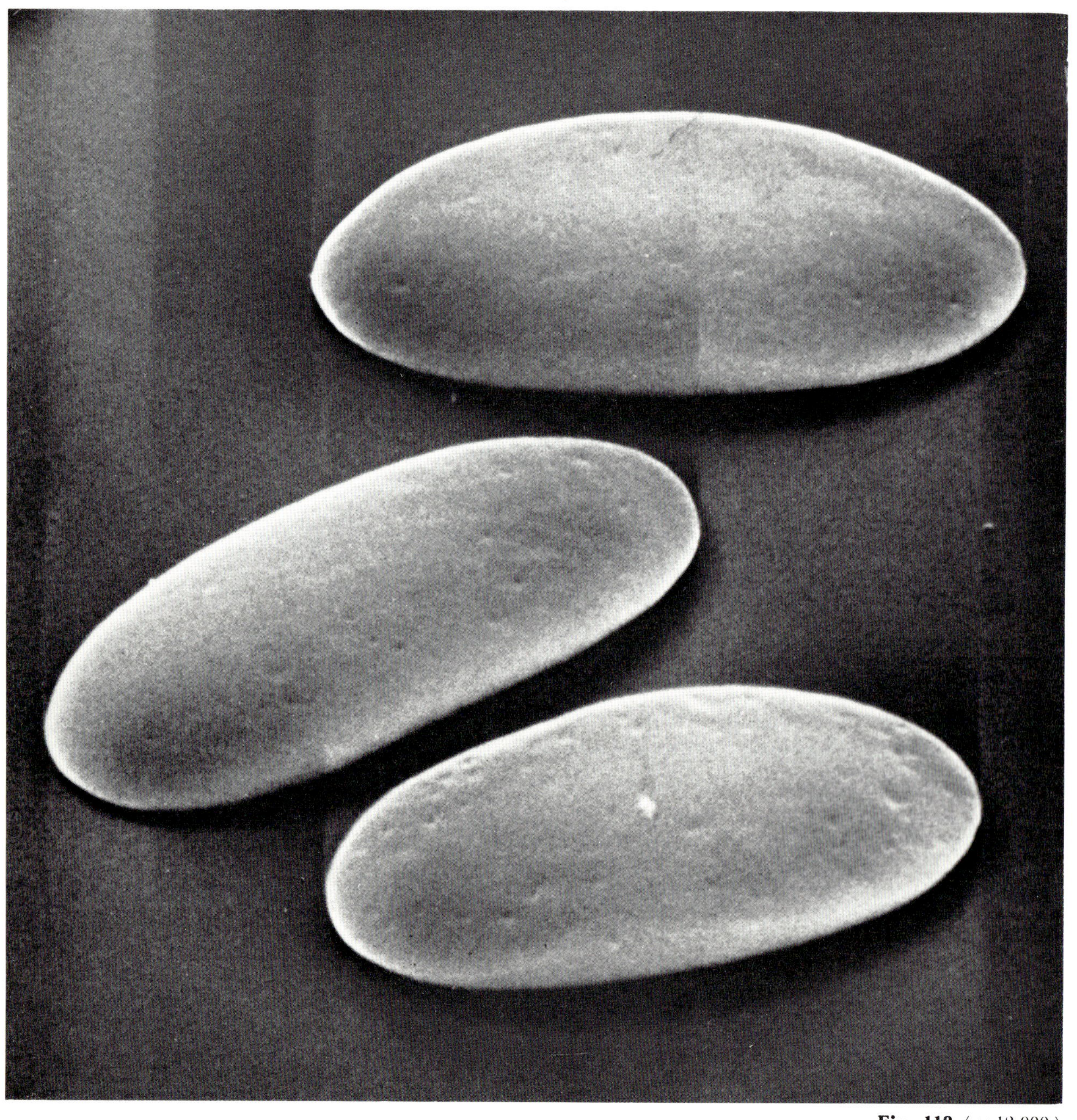

Fig. 118 (× 12,000)

Fig. 119 (× 12,000)

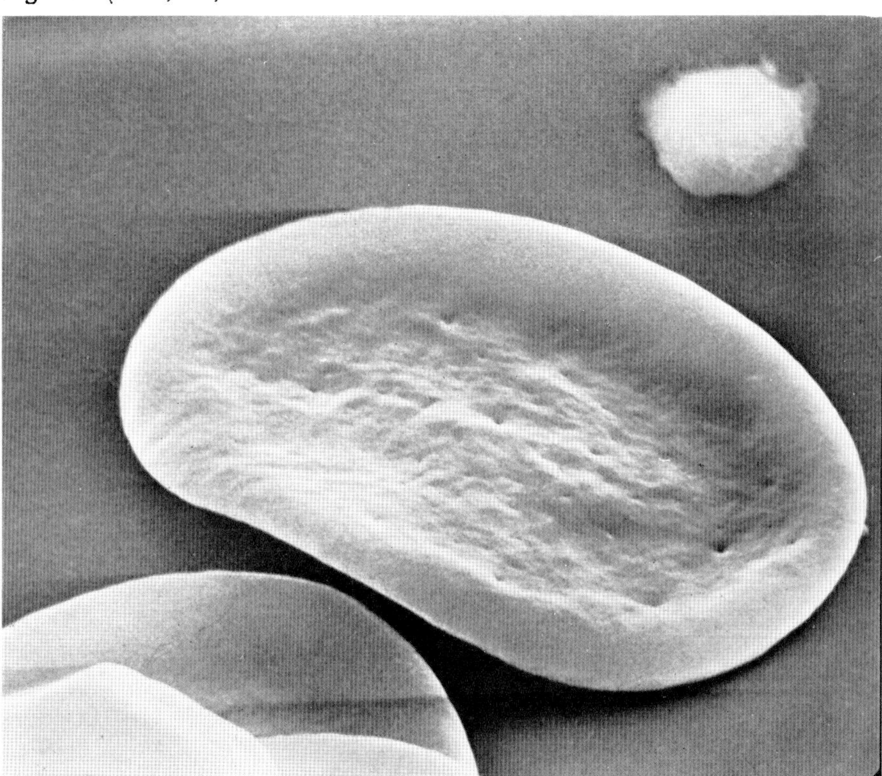

Fig. 119 shows the red cells of the toad which are also elliptic in shape but appear very thin with a slight thickening at the center corresponding to the site of the nucleus.

In both micrographs, especially in Fig. 119, one may clearly see tiny pits on the cell surface as if they were made by a pointed needle. The significance of these pits which also occur in some pathological red cells of man, is unknown.

Fig. 120 shows red cells of the camel. One may be surprised how they resemble the avian and amphibian cells shown in the preceding figures. The camel red cells, as in other mammals, are devoid of the nucleus but greatly deviate from the mammalian disc shape of red blood cells. The significance of this unique, so to speak atavistic character of camel red cells remains to be studied from physiological and phylogenetic points of view.

Note

The animals used were: an adult hen, an adult toad, *Bufo bufo japonicus* and an adult Bactrian camel. The specimen preparation was the same as shown on p. 75. EM: JSM-2 (Figs. 118 and 120) and HSM-2 (Fig. 119).

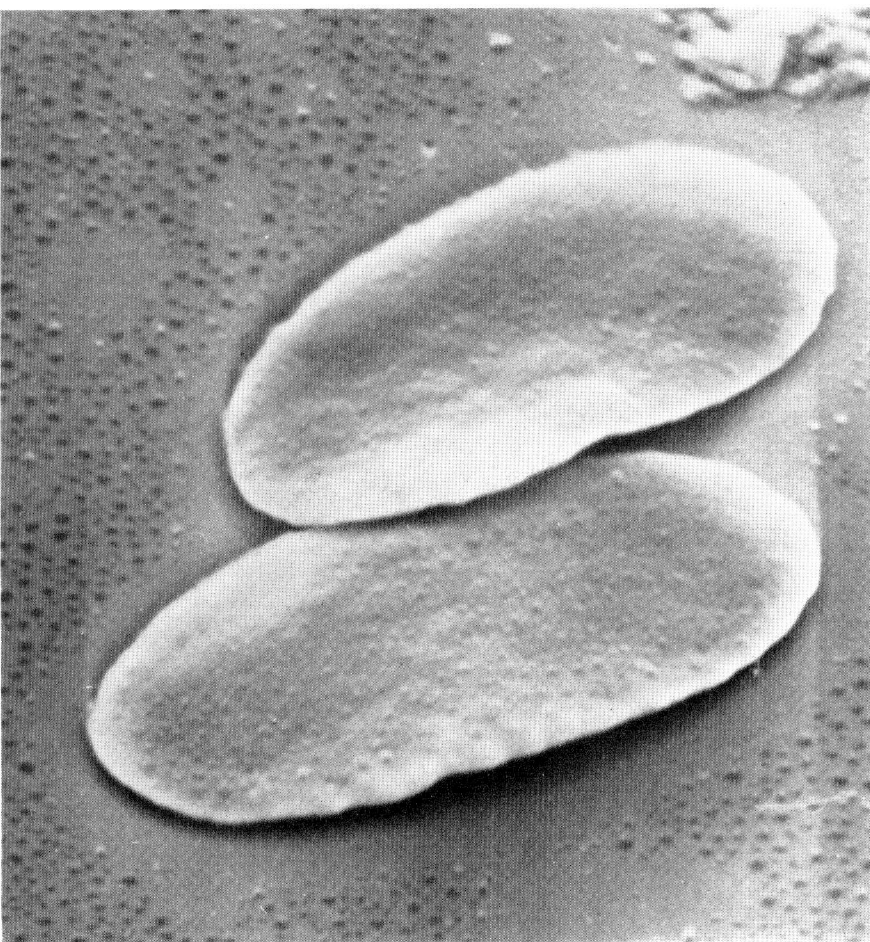

Fig. 120 (× 12,000)

30. Blood Platelets and their Transformations

Blood platelets are tiny but important elements of the blood, being involved in blood coagulation and hemostasis. They are too small in size to be precisely observed with the light microscope, whereas the transmission electron microscopy of sections is not an adequate methodology to analyse the subtle changes of their shape as a whole. Thus, the scanning electron microscopy promises fruit in the study of blood platelets.

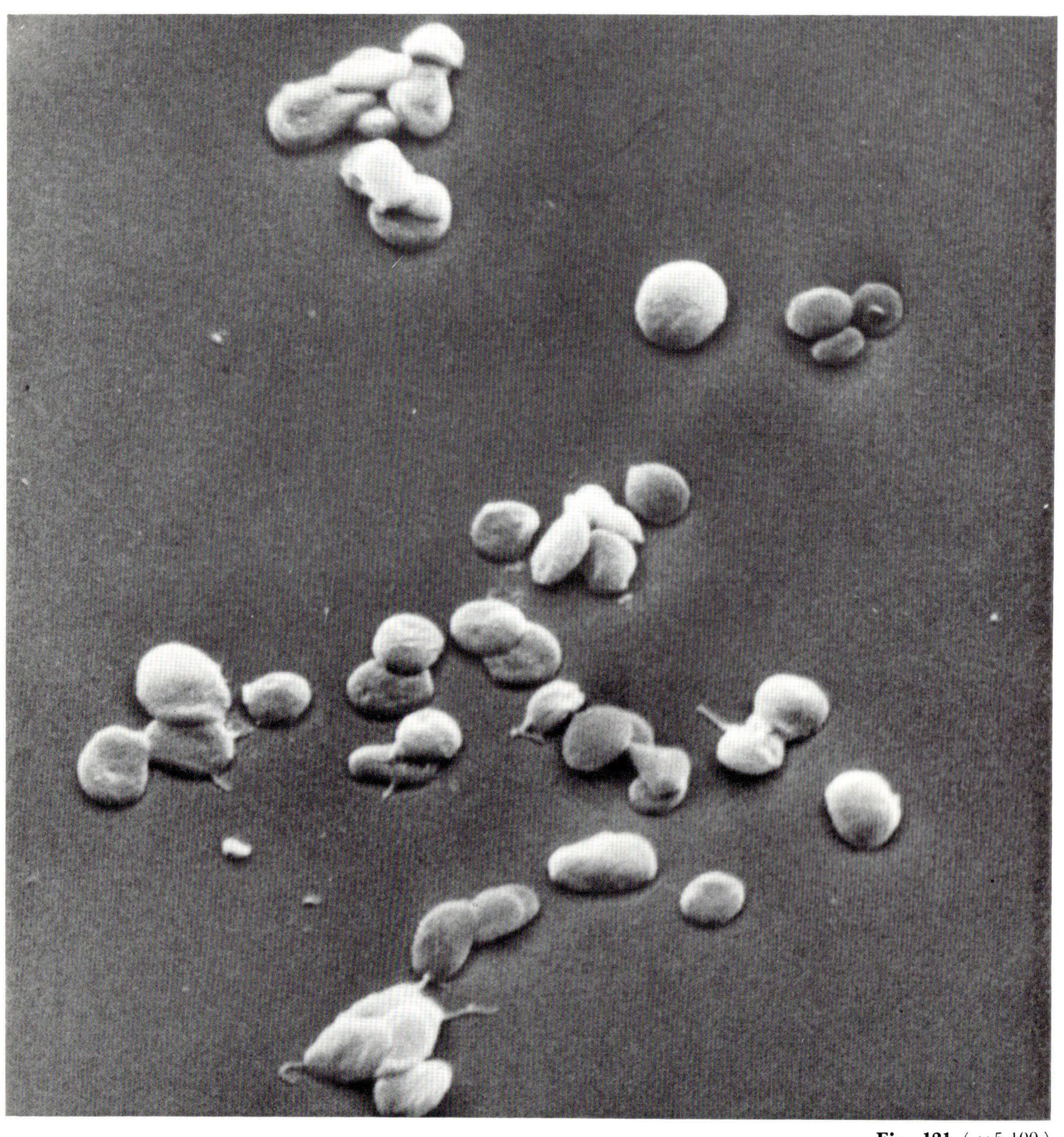

Fig. 121 (×5,100)

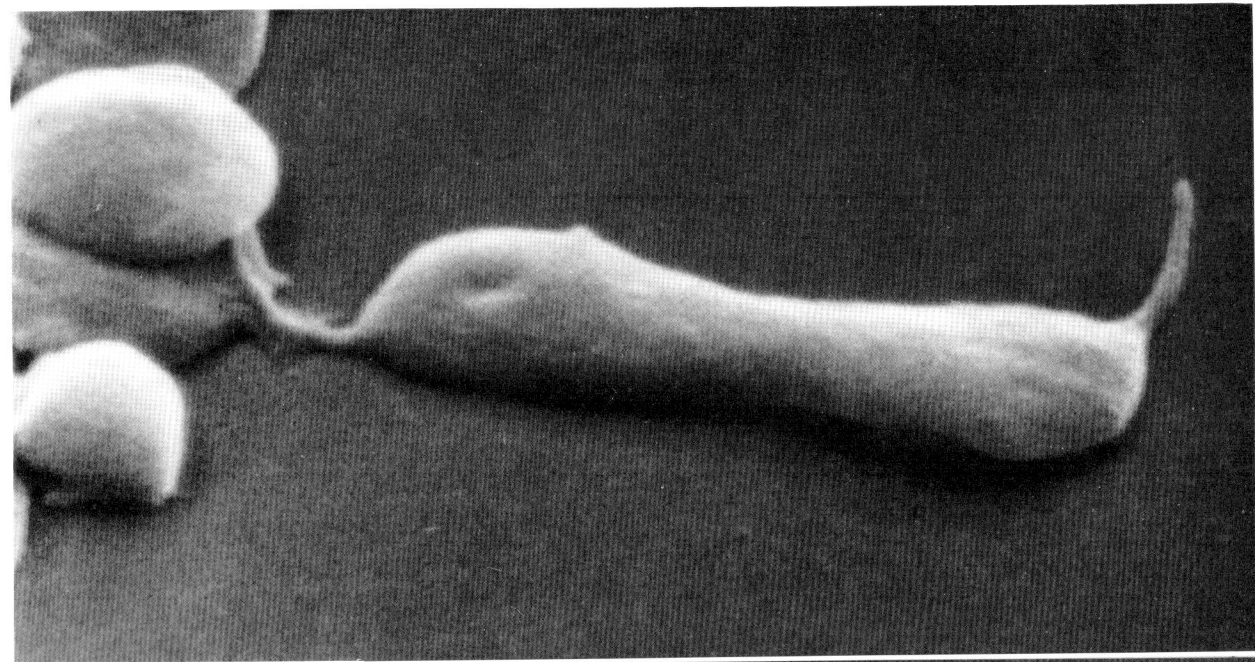

Fig. 122 (× 17,000)

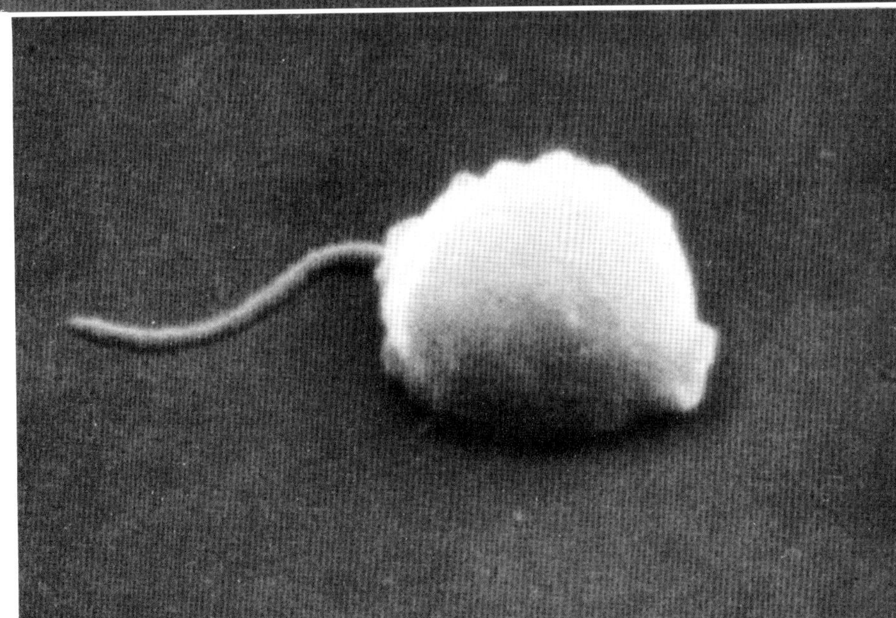

Fig. 123 (× 17,000)

Fig. 121 shows a population of normal human platelets which are mostly discoid or convex lens-like in shape. The platelets assume this form when the whole procedure of sampling has been performed very carefully so that physical and chemical stimuli upon them have been minimized. When the platelets are irritated, for instance when they are collected by less careful centrifugation, they change into thickened or irregular shapes and extend a few tentacle-like thin processes from their margin or equatorial side (**Figs. 122–124**). The surface of these platelets is smooth except for some dimples which probably correspond to the orifices of the endoplasmic reticulum which are known by transmission electron micrographs of sections.

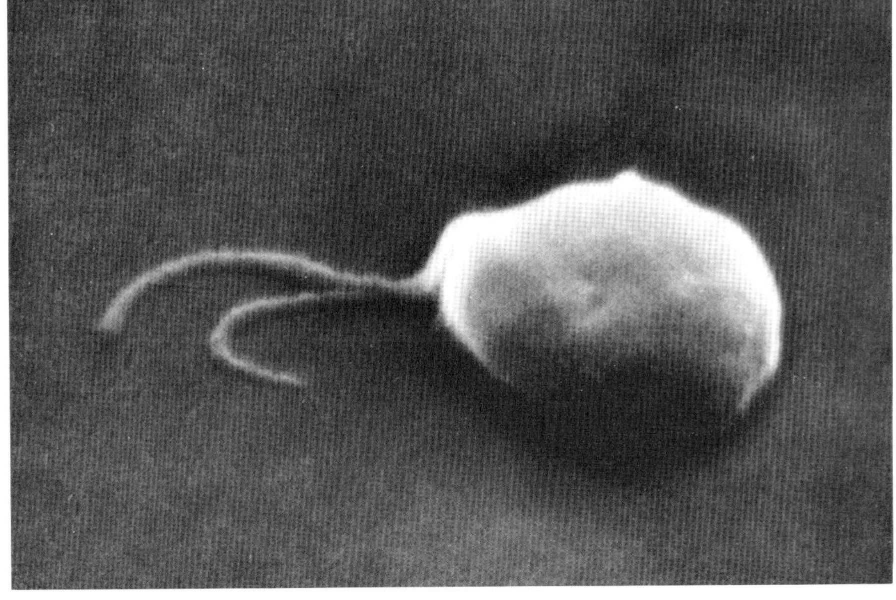

Fig. 124 (× 17,000)

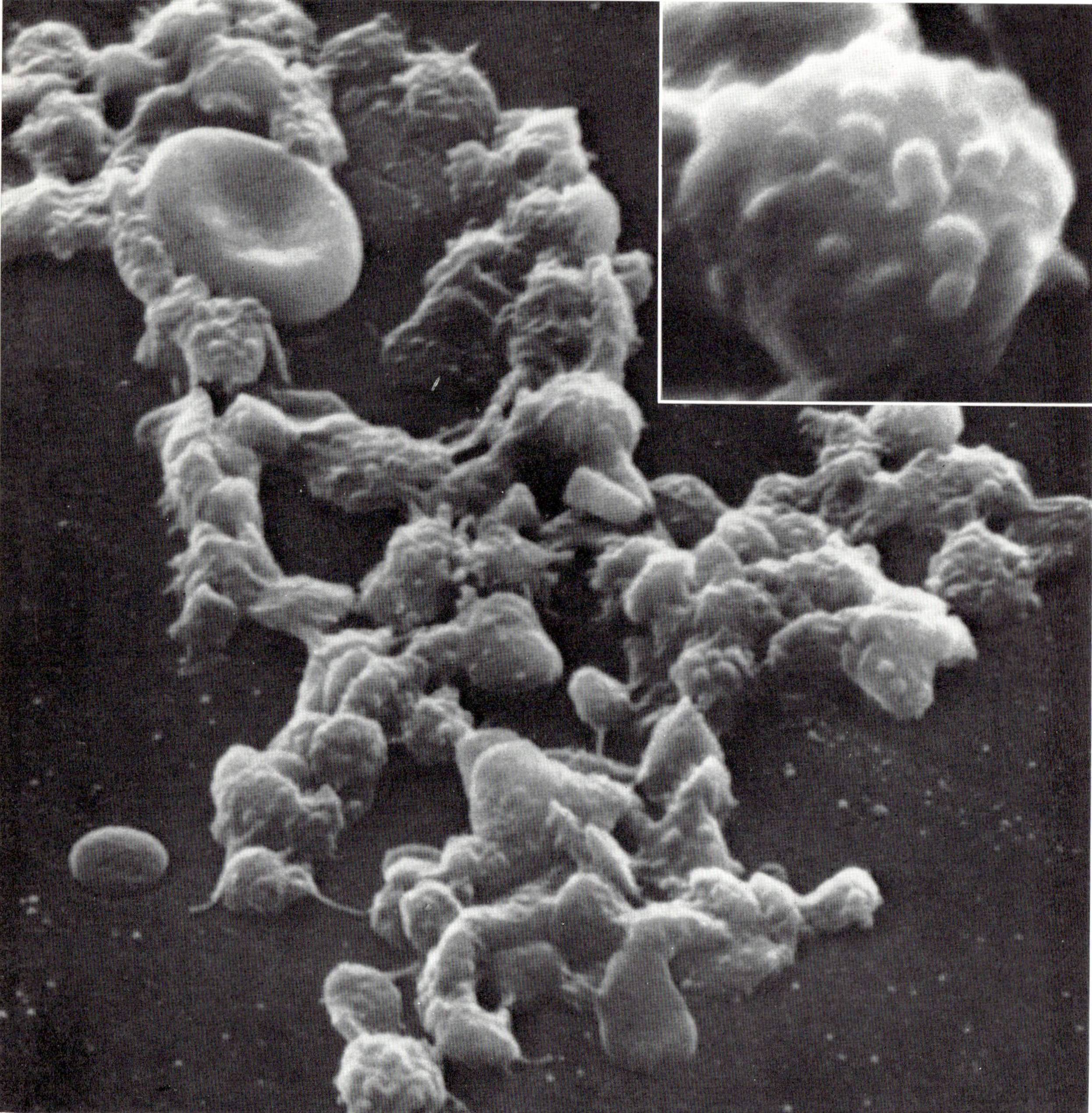

Fig. 126 (×36,000)

Fig. 125 (×6,800)

A series of conspicuous changes in the shape of platelets is caused when one adds thrombin (for doses see below) to the platelet-rich plasma. As is shown in **Figs. 125** and **126**, the platelets come to be aggregated and show thickening of their body and round projections on their surface within 5 seconds. The long, equatorial processes shown in Figs. 122–124 increase in number and thickness. **Fig. 128** is a close-up of one of the platelets of this change which reminds us of an imaginary life on some distant planet. In 15 seconds fibrin fibers appear around individual platelets and extend their network binding them (**Fig. 127**).

Note

Blood was taken from a normal adult male with a siliconized system using citrate as anticoagulant and centrifuged at 100g for 10 minutes. The platelet-rich supernatant was, after incubation for 30 minutes at 37°C, fixed in 1 per cent glutaraldehyde (0.1M phosphate buffer). In the thrombin experiments, 1 unit of bovine thrombin was added to the incubated specimen. The samples were dehydrated in acetone, dried on glass slides and coated with carbon and gold. EM: JSM-2

Fig. 127 (×17,000)

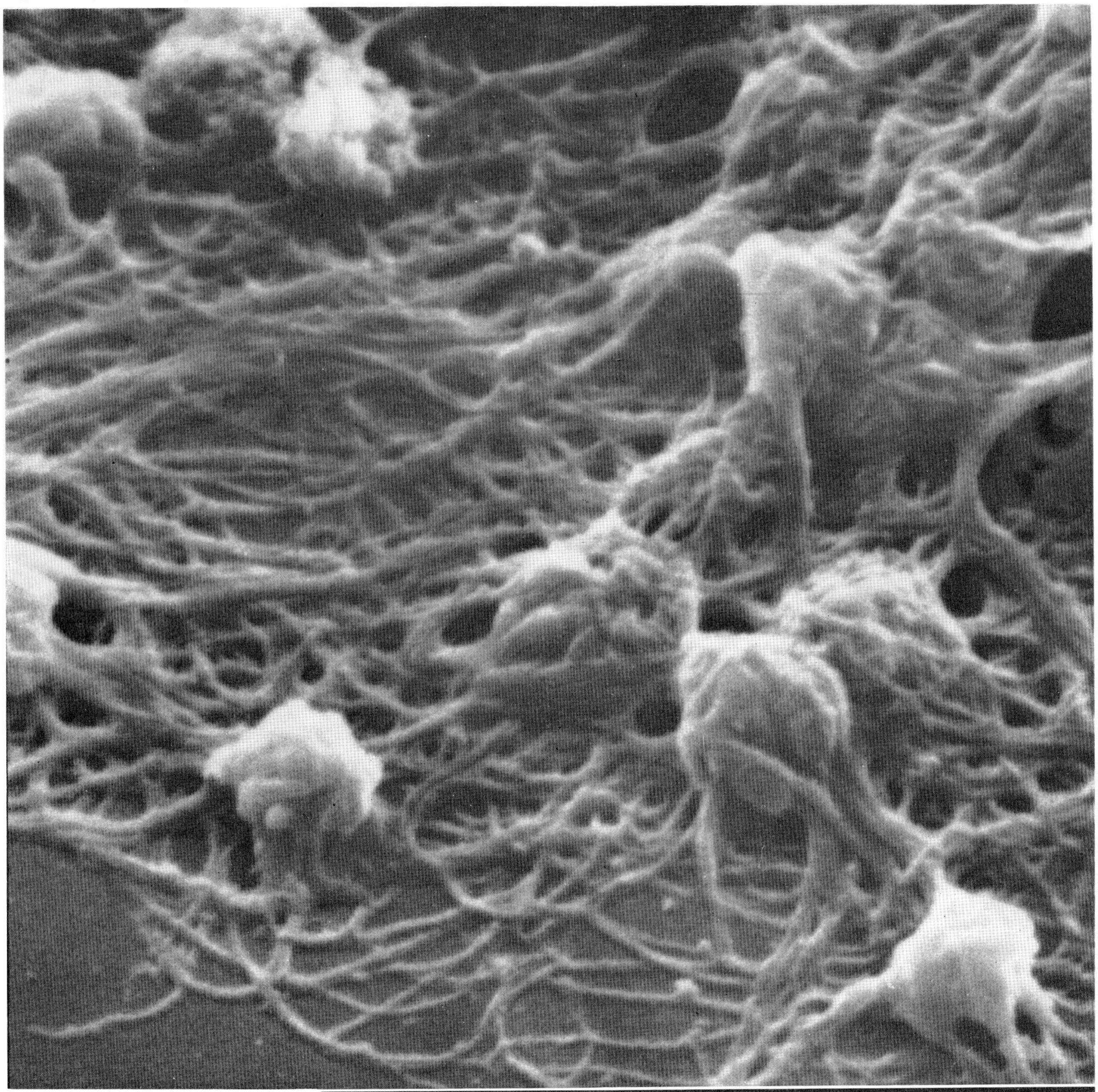

Reference

BARNHART, M. J. and J. M. RIDDLE: A three-dimensional study of platelet surface responses to drugs. Blood 34: 543 (1969).

CLARKE, J.A., C. HOWKEY and A.J. SALSBURY: Surface ultrastructure of platelets and thrombocytes. Nature 223: 401 (1969).

HATTORI, A., J. TOKUNAGA, T. FUJITA and M. MATSUOKA: Scanning electron microscopic observations on human blood platelets and their alterations induced by thrombin. Arch. histol. jap. 31: 37-54 (1969).

Figs. 122–128 by courtesy of the Arch. histol. jap.

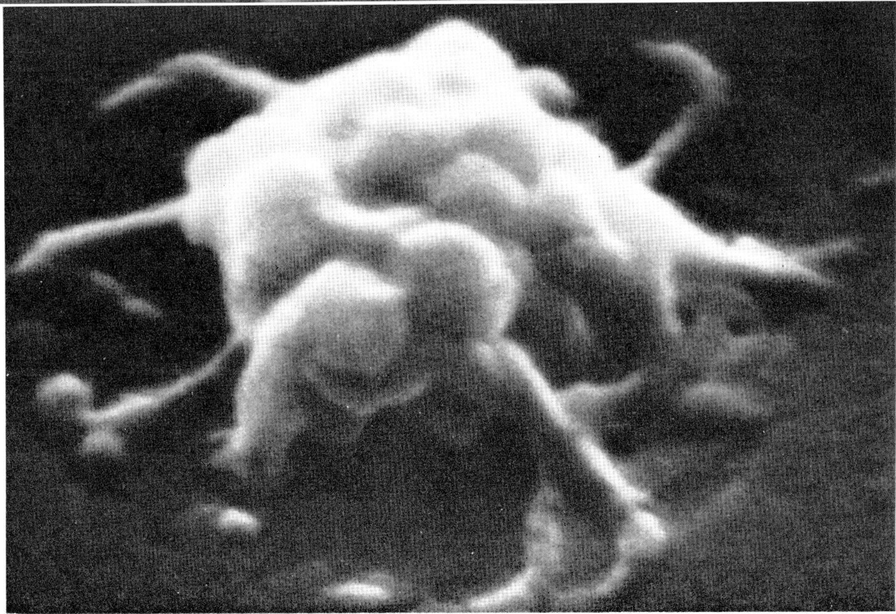

Fig. 128 (×36,000)

31. Spermatozoa of Normal and Sterile Men

The complicated interior structures of spermatozoa have been elucidated by the transmission electron microscopy of sections, but it has been difficult for these winding thread forms to be examined in their entire shape by this method. Even more difficult is the examination of the deformities of the spermatozoa which are examined daily with the light microscope in the urological clinic and are important for the study and diagnosis of male sterility. The advent of the scanning electron microscope is thus believed epochal for studies along this line. The difficulty of eliminating the mucous sperm plasma of the ejaculate has been overcome by the present authors.

Figs. 129–131 show spermatozoa from an ejaculate of a normal man. The characteristic racket shape of the head and the long tail reaching 60μ are obvious in **Fig. 129.** **Fig. 130** is a close-up of a part of Fig. 129 in which two heads are seen in top and side views. The anterior portion of the head is armed with a thick hood called

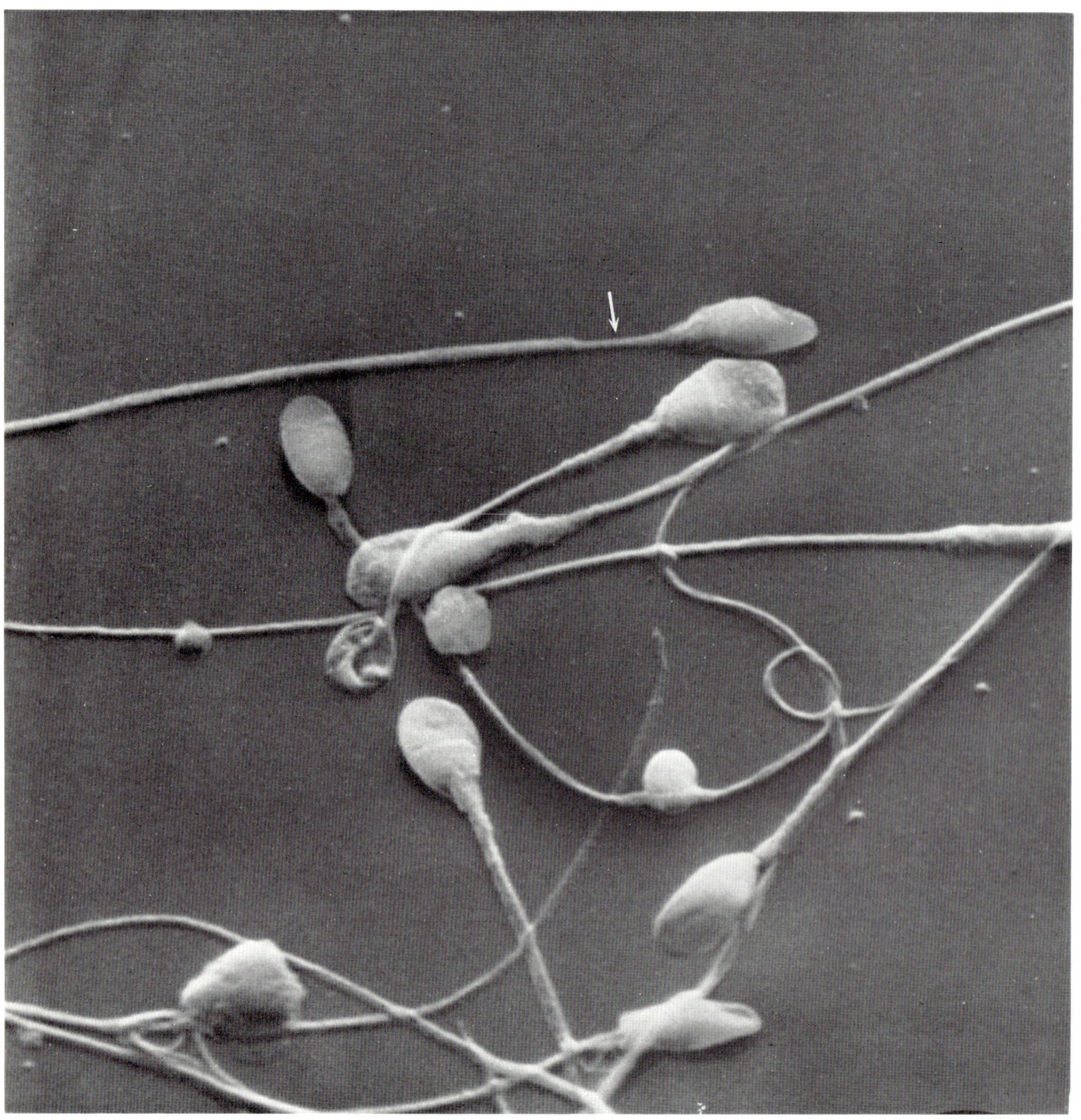

Fig. 129 (×5,100)

Fig. **130** (× 12,000)

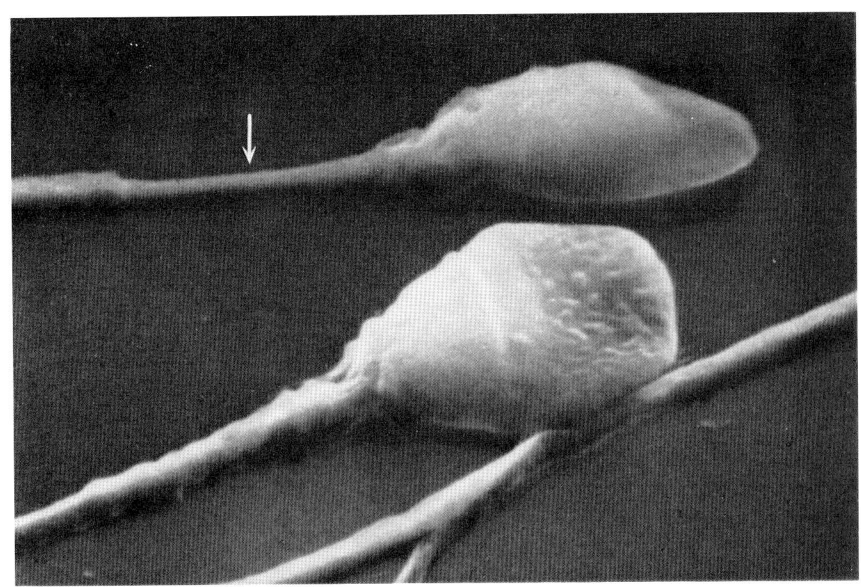

Fig. **131** (× 12,000)

acrosome while the posterior tapering portion has a thinner sheath called postnuclear cap. A transverse furrow marks a space between them. The middle piece of the tail just behind the head shows an undulating relief. This corresponds to mitochondria which spiral around this portion of the tail.

In a spermatozoon shown in **Fig. 129** and, in a closer view, in **Fig. 130** this mitochondrial thickening is lacking (arrow). This deformed spermatozoon deprived of its "motor" would have been immotile in the ejaculate.

The often constricted boundary of the middle and main pieces of the tail likely represents a mechanically weak point as suggested in **Fig. 131** by a spermatozoon broken at this site. Fig. 131 also indicates the main piece abruptly going to the short end-piece (arrow).

Fig. 132 (× 10,000)

Fig. 132 shows a spermatozoon with two heads found in the same normal semen. The acrosome is likely failing in these heads.

Although deformities are found in no more than 15 per cent of the total spermatozoa in normal semen, they occur more or less in excess or even exclusively in the ejaculates of sterile males. **Figs. 133–135** are abnormal spermatozoa from patients who came to the clinic because of suspected male sterility. The spermatozoon in **Fig. 133** has a monstrous head and a short and irregularly thick tail. **Fig. 134** indicates a form with a rudimental and bipartite head and double tails. The one in **Fig. 135** has an unusually small, oval and rough-surfaced head without an acrosome. All the forms shown in these three figures are of immotile type, being devoid of a mitochondrial thickening in the middle piece.

The specimens of sterile semen were provided by the Department of Urology of Niigata University School of Medicine.

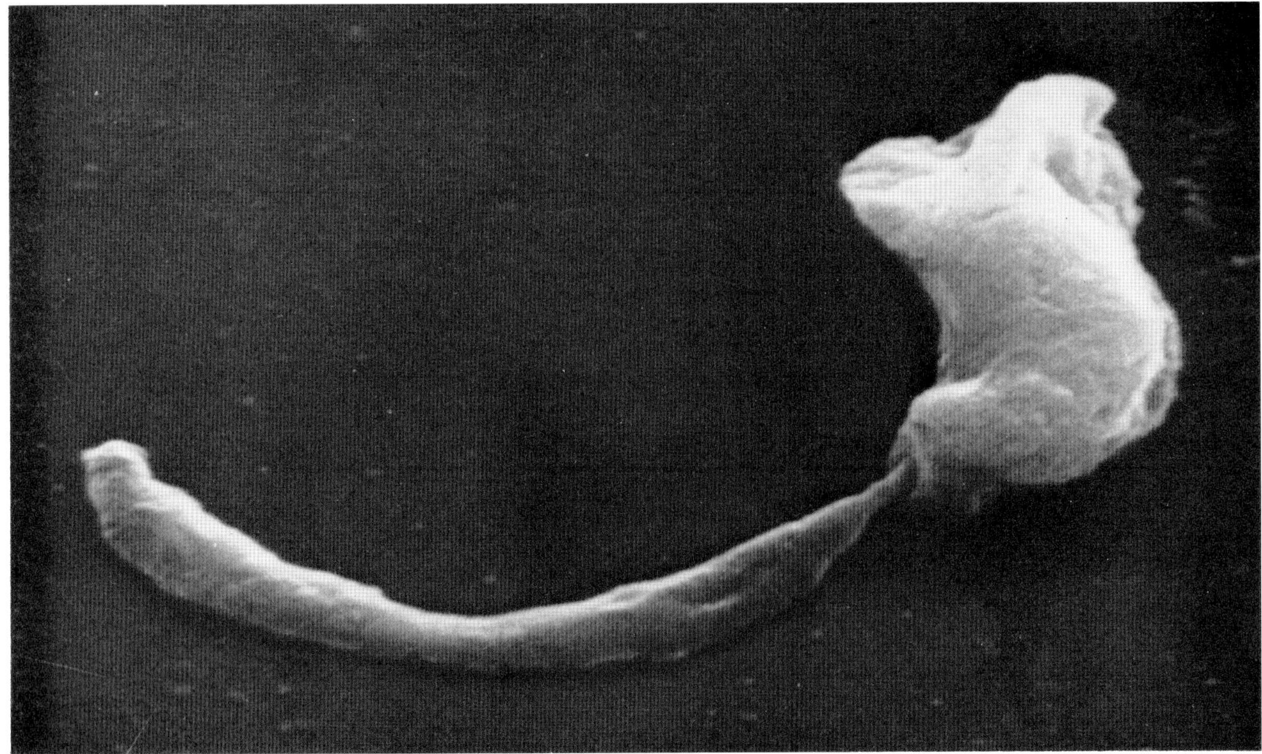

Fig. 133 (× 10,000)

Fig. 134 (× 8,700)

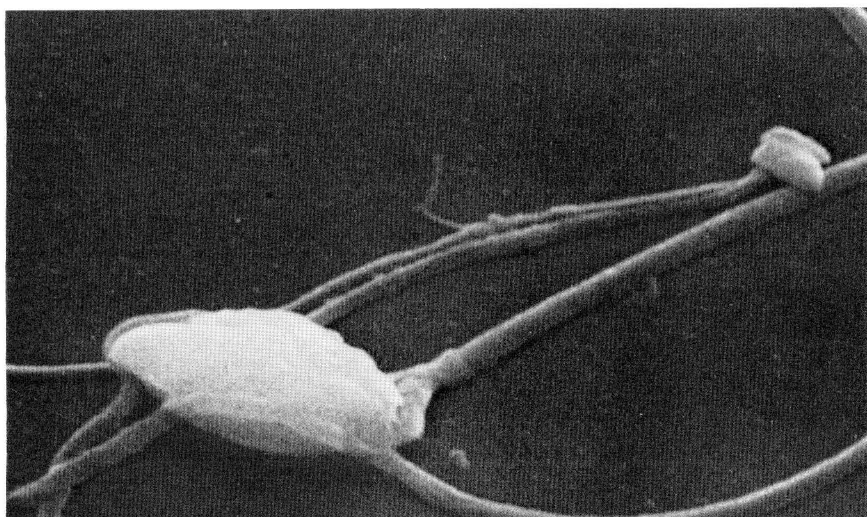

Note

Ejaculates after liquification were fixed for one hour either in 10 per cent formalin (Figs. 129 and 130) or in 1 per cent glutaraldehyde in 0.1 M phosphate buffer (Figs. 131-135) according to the method described on p. 2. Specimens were dehydrated in acetone, dropped on glass slides to be dried in air, and coated with carbon and gold. EM: JSM-2 (Figs. 129-132) and HSM-2 (Figs. 133-135)

Reference

FUJITA, T., M. MIYOSHI and J. TOKUNAGA: Scanning and transmission electron microscopy of human ejaculate spermatozoa with special reference to their abnormal forms. Z. Zellforsch. 105: 483-497 (1970).

Figs. 130-135 by courtesy of the Editor of the Z. Zellforschung and the Springer Verlag.

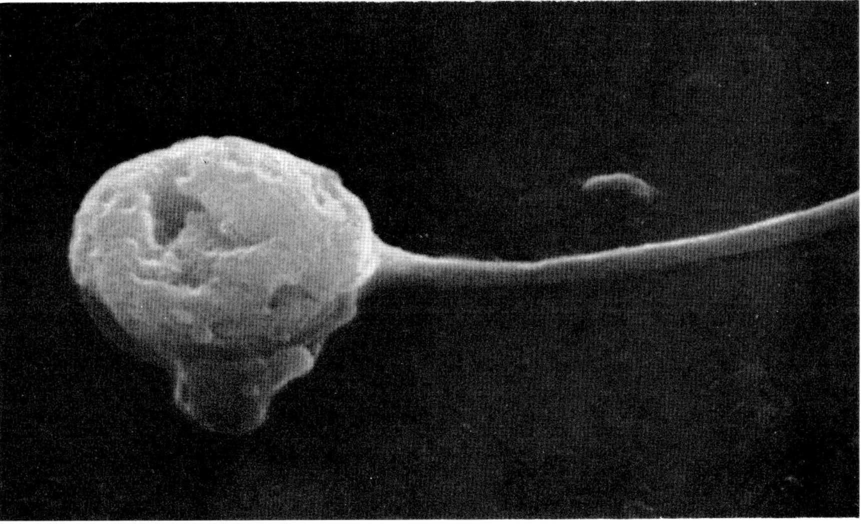

Fig. 135 (× 20,000)

32. Joint Fluid in Rheumatoid Arthritis and Gout

Although joint fluid or synovia in the normal joint contains some white blood cells and debris of cellular and fibrillar elements of the joint wall, pathological conditions in the joint may cause the occurrence of unusual forms which can be impressively demonstrated by scanning electron microscopy. In this section two conspicuous instances will be shown.

Fig. 136 is the joint fluid taken from a rheumatoid knee joint of a woman. Numerous free cells of round shape are almost exclusively neutrophil leucocytes as was confirmed by light microscopic examination of stained smears of the same fluid.

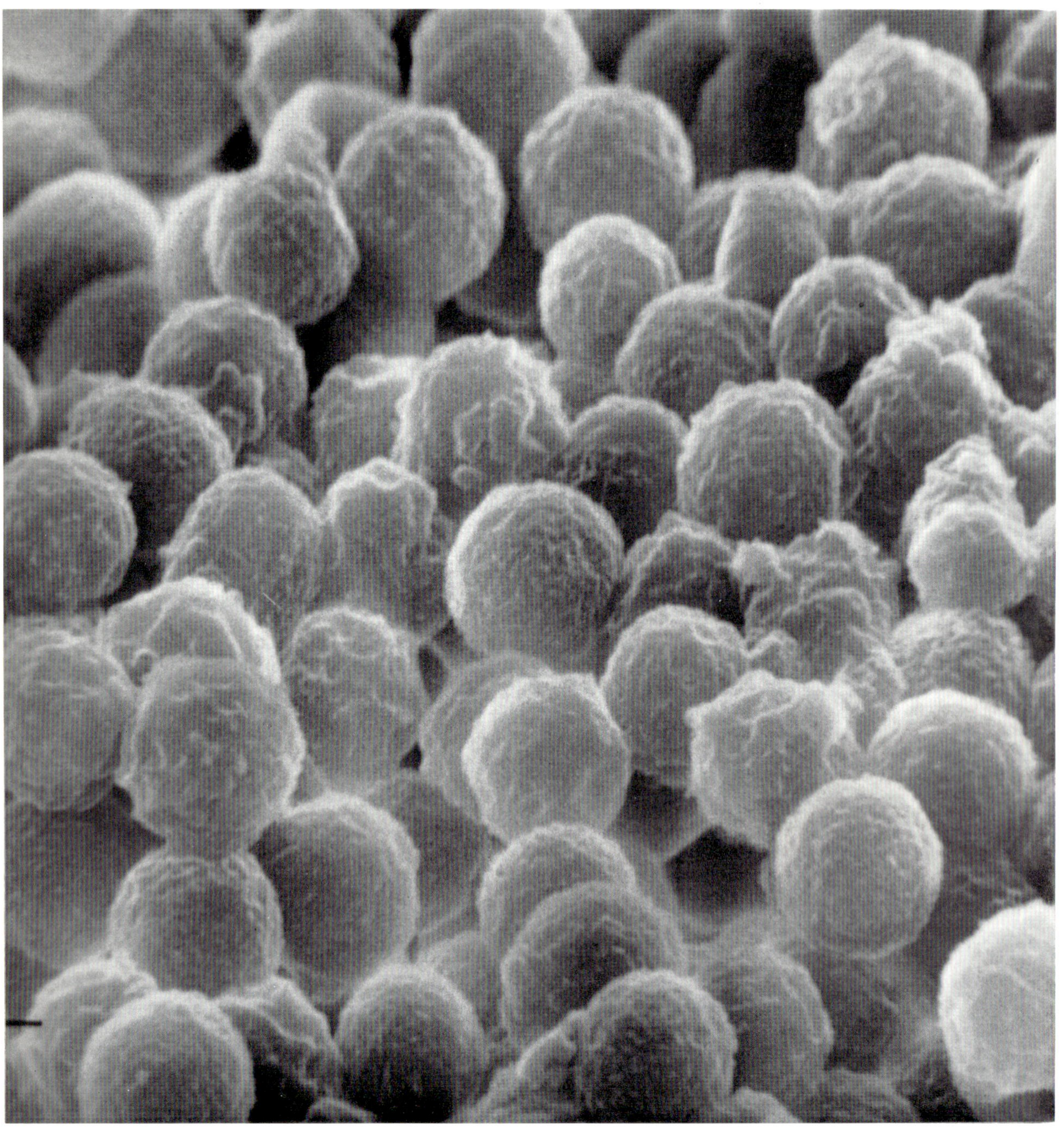

Fig. 136 (×5,100)

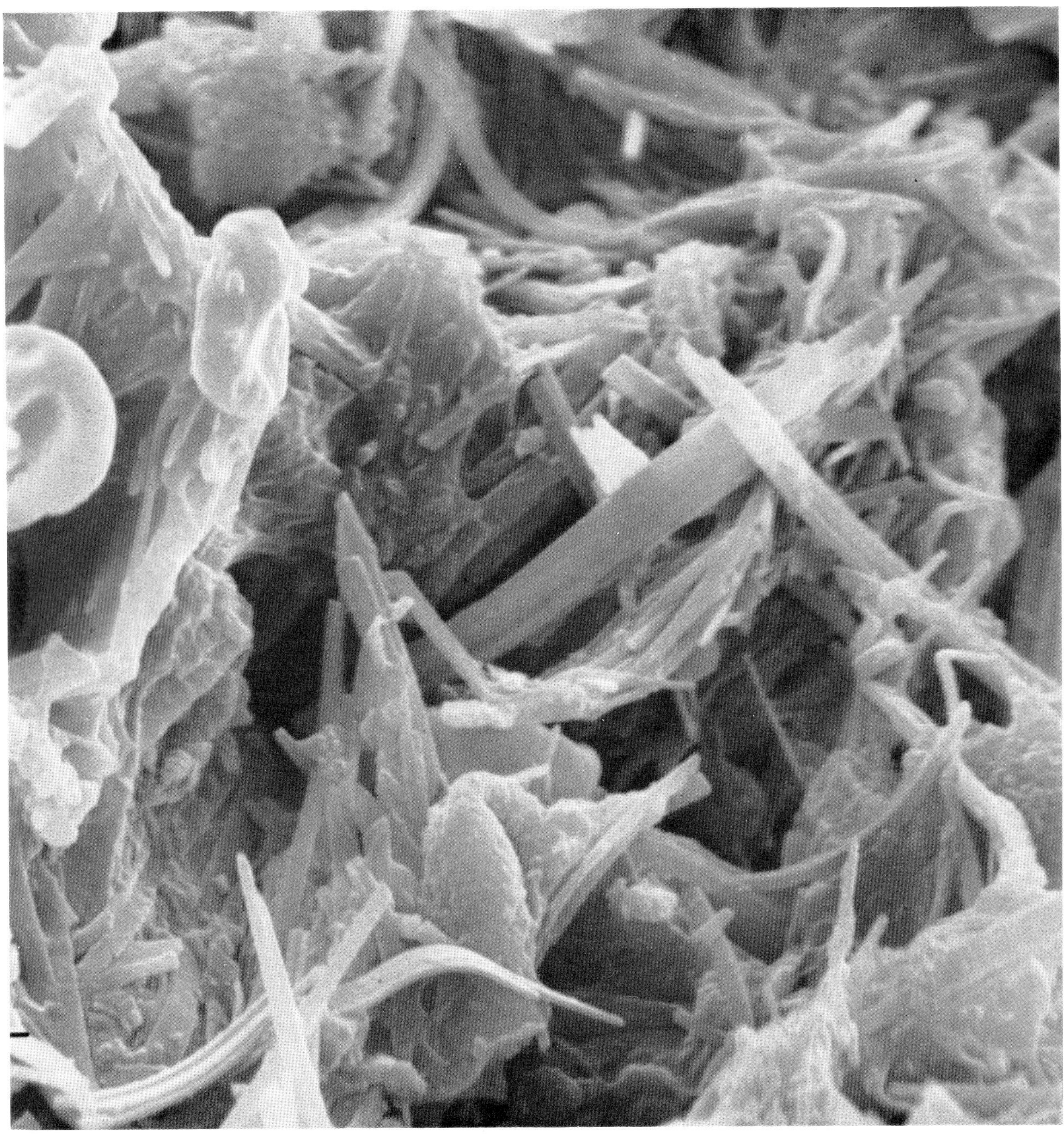

Fig. 137 (×5,100)

In most cases of rheumatoid arthritis a conspicuous number of free cells are found in the joint fluid, neutrophil leucocytes being an overwhelming majority. They may have a count as high as 10,000/cmm. The possible role of these leucocytes in the autoimmune mechanism in rheumatoid arthritis has been much disputed.

Fig. 137 shows the fluid taken from a swollen joint of the toe of a 55-year-old man suffering from gout. The abundant beam-shaped forms which partly appear bent are crystallized uric acid. A few erythrocytes and other cells are mingled with them.

Gout is a disease characterized by a high uric acid content in the serum and by painful joints which contain uric acid crystals. Detection of the crystals with the polarization microscope has been critical in the diagnosis of this disease.

Note

Joint fluid taken by puncture was fixed in 2.5 per cent glutaraldehyde (0.1M phosphate buffer), dehydrated in acetone, dropped on glass slides to be dried in air, and coated with carbon and gold. EM: JSM-2

CELL ORGANELLES

33. Isolated Mitochondria

Scanning electron microscopy of subcellular fractions has two purposes. The one is to understand the structures and the changes of the isolated cell organelles in their three-dimensional entity, while the other is to check, simply and quickly, the results of fractionation.

In 1969 our research group obtained the first convincing scanning electron micrographs of isolated mitochondria (Kurahasi et al., 1969), and these micrographs are reproduced here. These figures indicate variable forms of mitochondria according to the concentrations of sucrose used in fractionation.

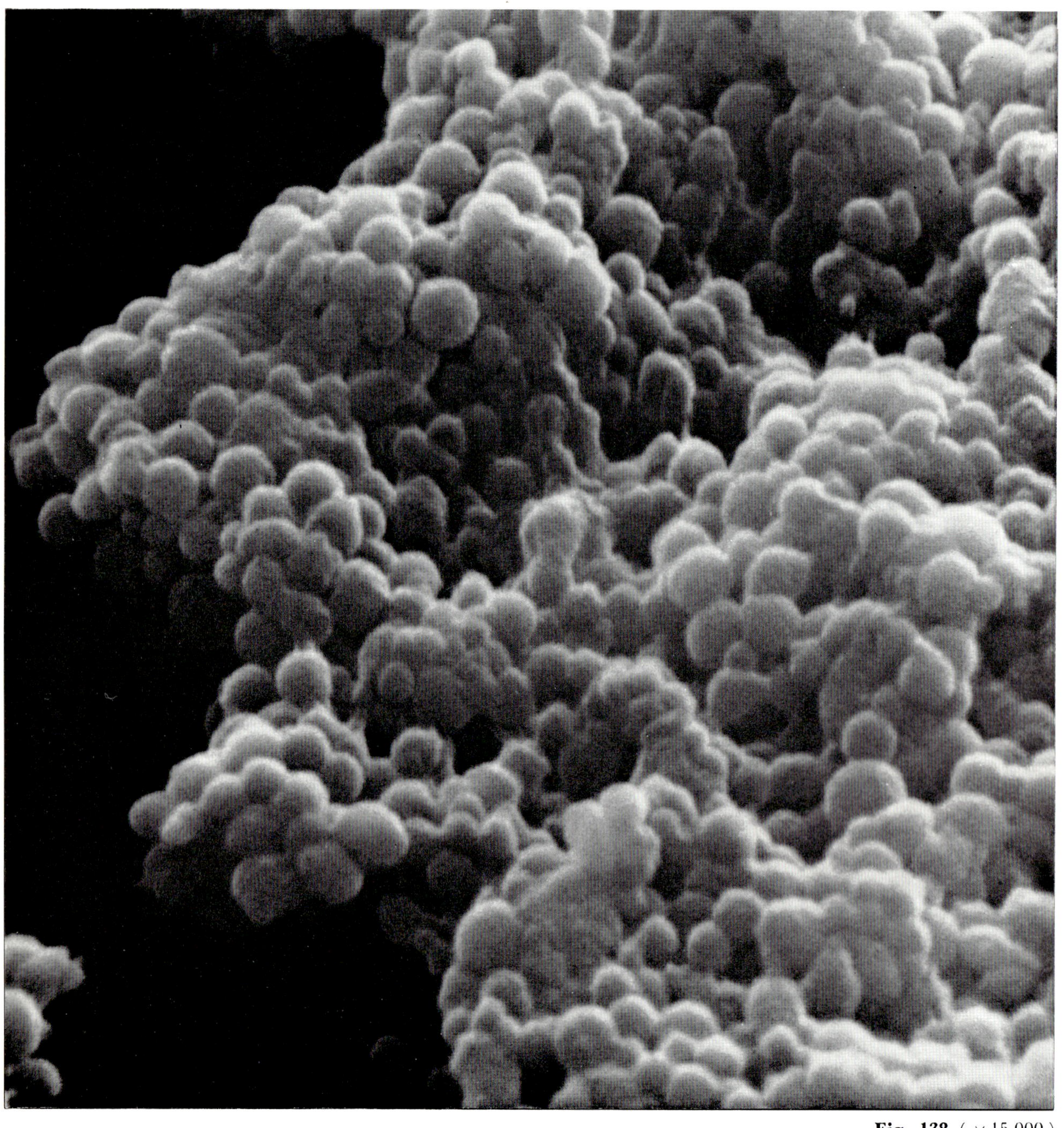

Fig. 138 (× 15,000)

Fig. 139 ($\times$36,000)

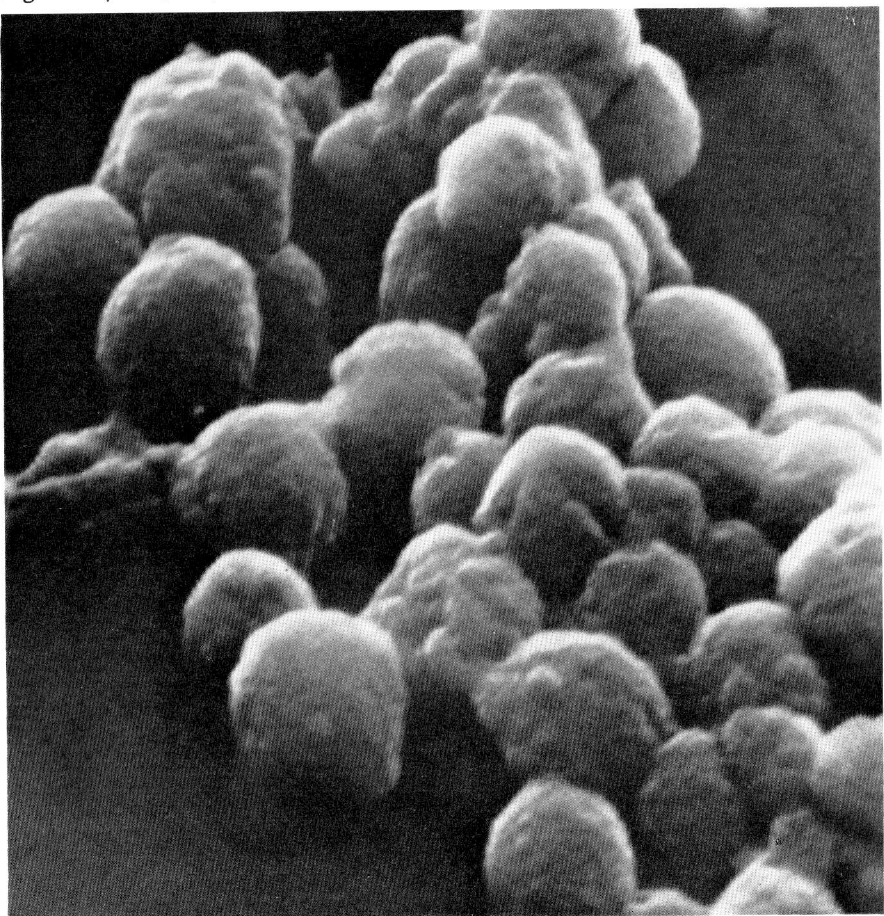

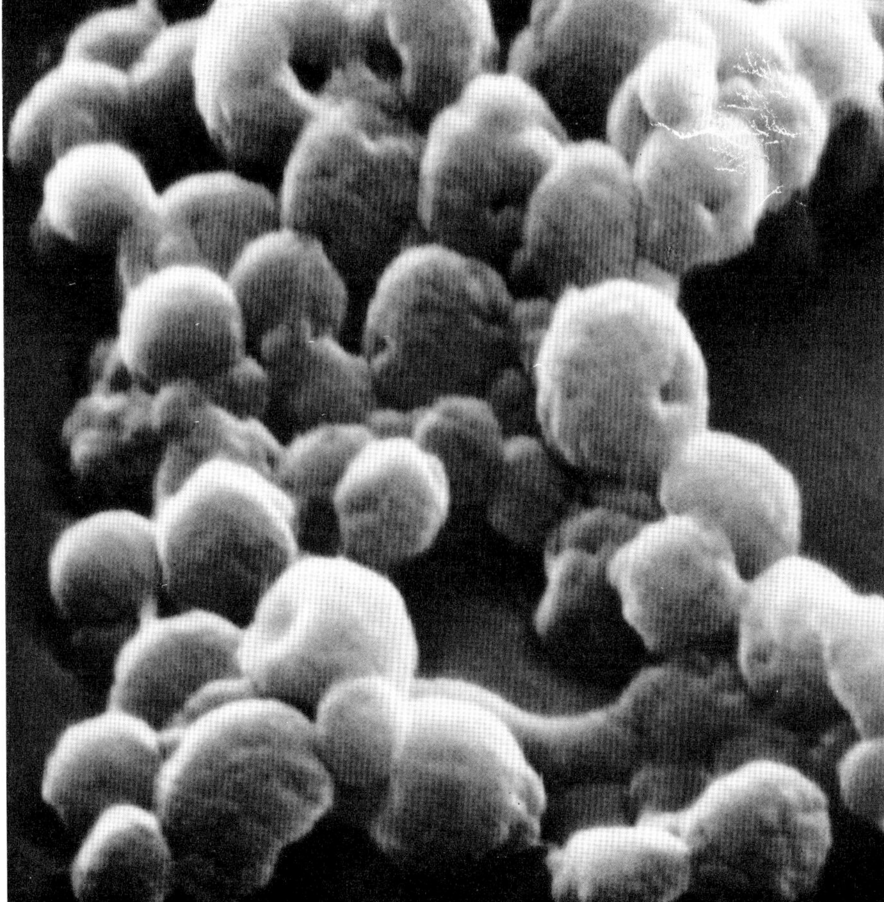

Fig. 140 ($\times$36,000)

Fig. 138 shows mitochondria isolated from the rat liver using, as in routine technique, 0.25M sucrose. As a high-power micrograph (**Fig. 139**) indicates, they are generally rounded in shape, with lumpy elevations on the surface. Their size is variable, mostly falling in the range of 0.4 to 0.7μ in diameter.

When the fractionation is made in a little higher osmotic pressure, that is, in 0.32M sucrose, there occur, besides apparently simple spheres, spherical bodies with a deep bore and, moreover, ring forms resembling doughnuts. **Fig. 140** indicates all of these forms and even a rod-shaped mitochondrion at the bottom of the picture.

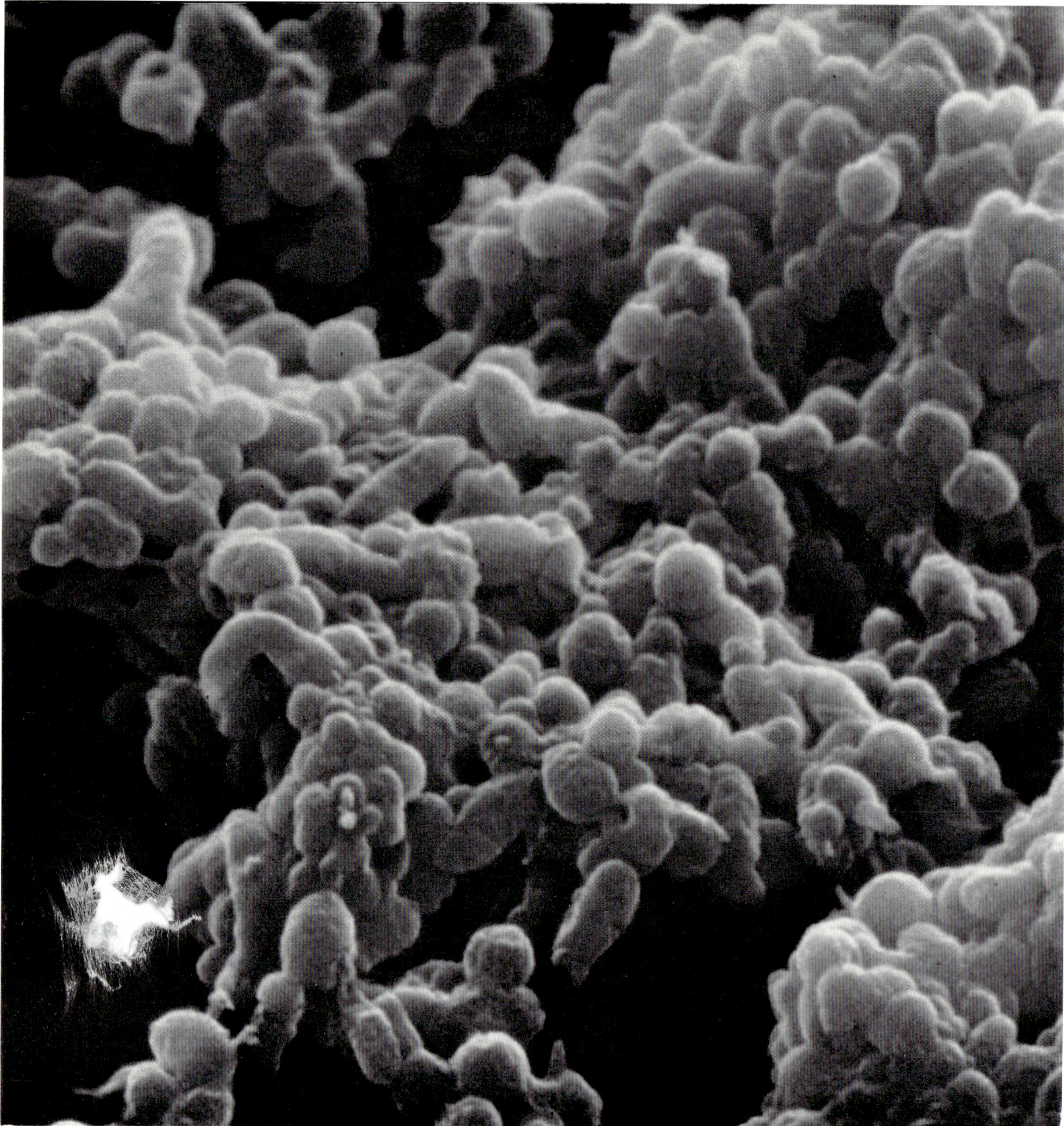

Fig. 141 (×17,000)

When hypertonic sucrose of 0.44M concentration is used in fractionation, numerous rod-like mitochondria appear as shown in **Fig. 141.** There are mingled also U-shaped and doughnut-like ones. Closer views of these forms are shown in **Figs. 142** and **143**. Notches occurring on the mitochondrial surfaces are thought to have been caused by the hypertonic effect of the medium.

It seems thus apparent that the rod shape of mitochondria in a hypertonic medium rolls up into U-shape and, by a fusion of both ends, into a ring when the osmotic pressure of the medium is lowered, and in a standard 0.25M sucrose solution, it is transformed into a sphere by a closure of the central bore.

Fig. 142 (×48,000)

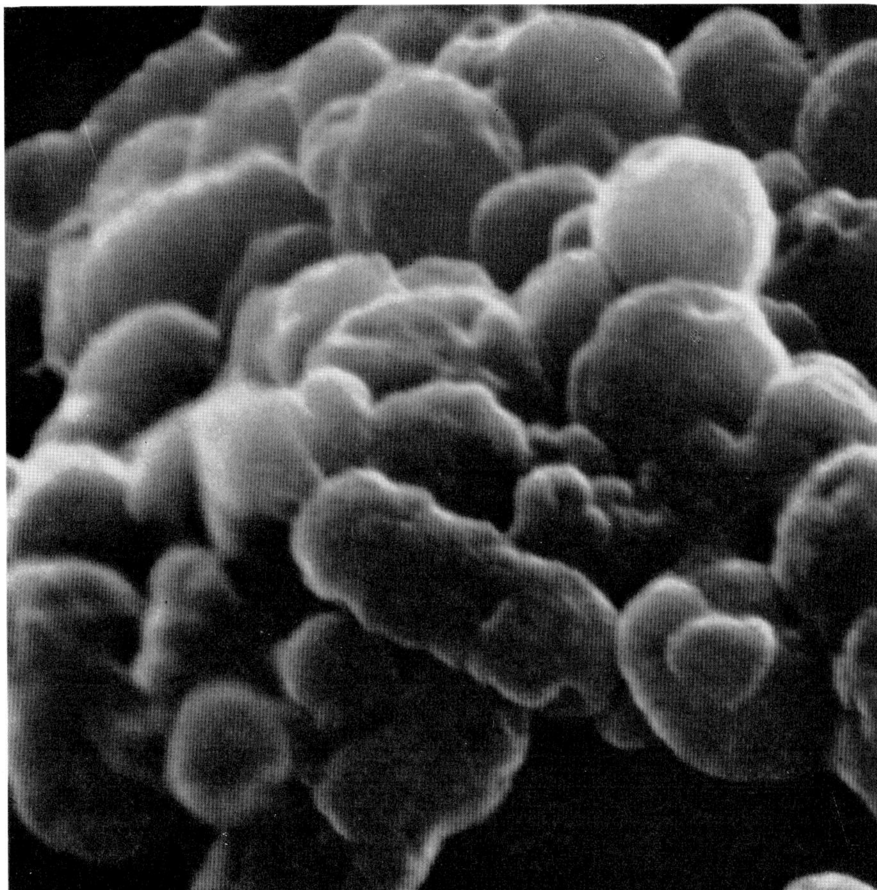

The variable forms of mitochondria under the scanning electron microscope remind us of some worms bending and stretching their bodies. It may be interesting to recollect that mitochondria in living cells may actively change back and forth between the thread type ("mitos") and the granule type ("chondros").

These findings were obtained in collaboration with Dr. KENGO KURAHASI, Department of Biochemistry, Okayama University School of Medicine.

Note

The rat liver was homogenized in an ice-cold 0.25 M, 0.32 M or 0.44 M sucrose solution. After nuclei and cell debris were precipitated, the supernatant was centrifuged at 6,000 × g, 8,000 × g or 12,000 × g respectively. Mitochondria thus isolated were suspended for 20 minutes in 2.5 per cent glutaraldehyde in a 5mM Tris buffer (pH 7.4) containing 20mM KCl, 5mM succinate and sucrose of the same concentration as that used in fractionation. After dehydration in acetone, a suspension of mitochondria in absolute acetone was dropped on glass slides to be dried in air, and was coated with carbon and gold. EM: JSM-2

Reference

KURAHASI, K., J. TOKUNAGA, T. FUJITA and M. MIYAHARA: Scanning electron microscopy of isolated mitochondria. I. Arch. histol. jap. 30: 217-232 (1969).

Figs. 138–143 by courtesy of the Arch. histol. jap.

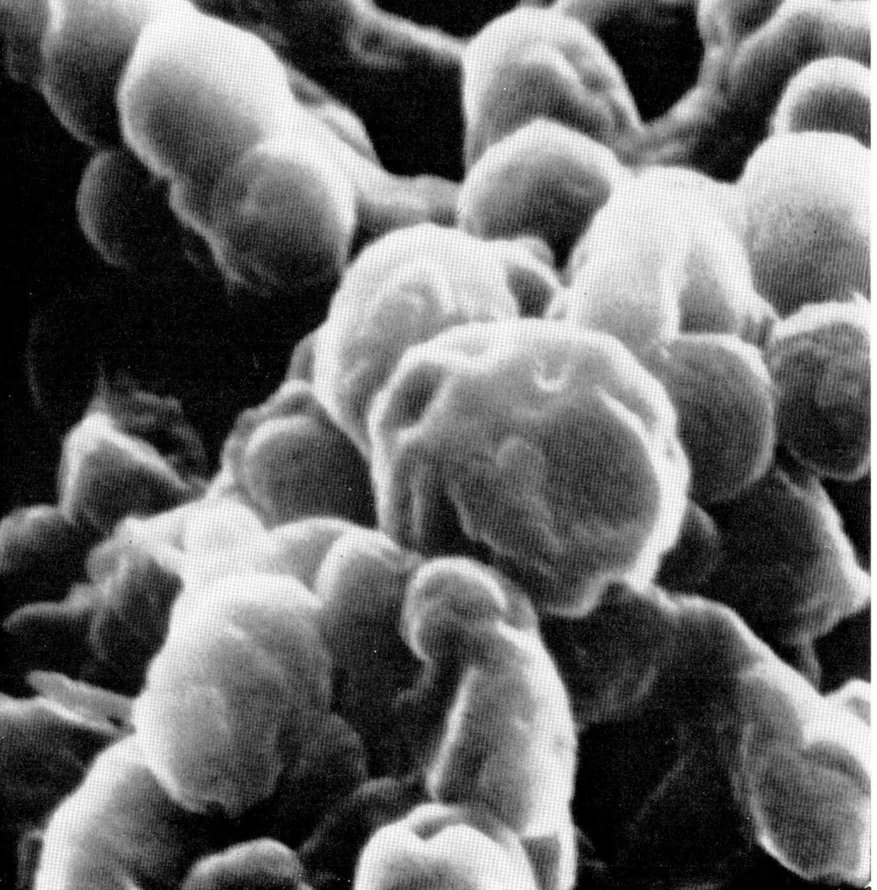

Fig. 143 (×48,000)

34. Human Chromosomes

Recent progress in karyoanalysis has elucidated the relationships between various types of congenital disorders and the abnormalities in chromosomes. Although only light microscopy has been used thus far for studies along this line, it is now expected that scanning electron microscopy will also be applied to this field and will serve to elucidate the changes of chromosomes at the submicroscopic level.

Fig. 144 is a scanning electron micrograph of a normal set of metaphase chromosomes obtained from the white blood cells of a healthy male showing 44 plus X and Y chromosomes. **Fig. 145** is a closer view of a part of Fig. 144. In some chromosomes one may recognize a spiral structure. **Fig. 146** shows one of the No.1 chromosomes at a high magnification. At the site indicated by the arrow, the double coiling of a chromosome may be recognized.

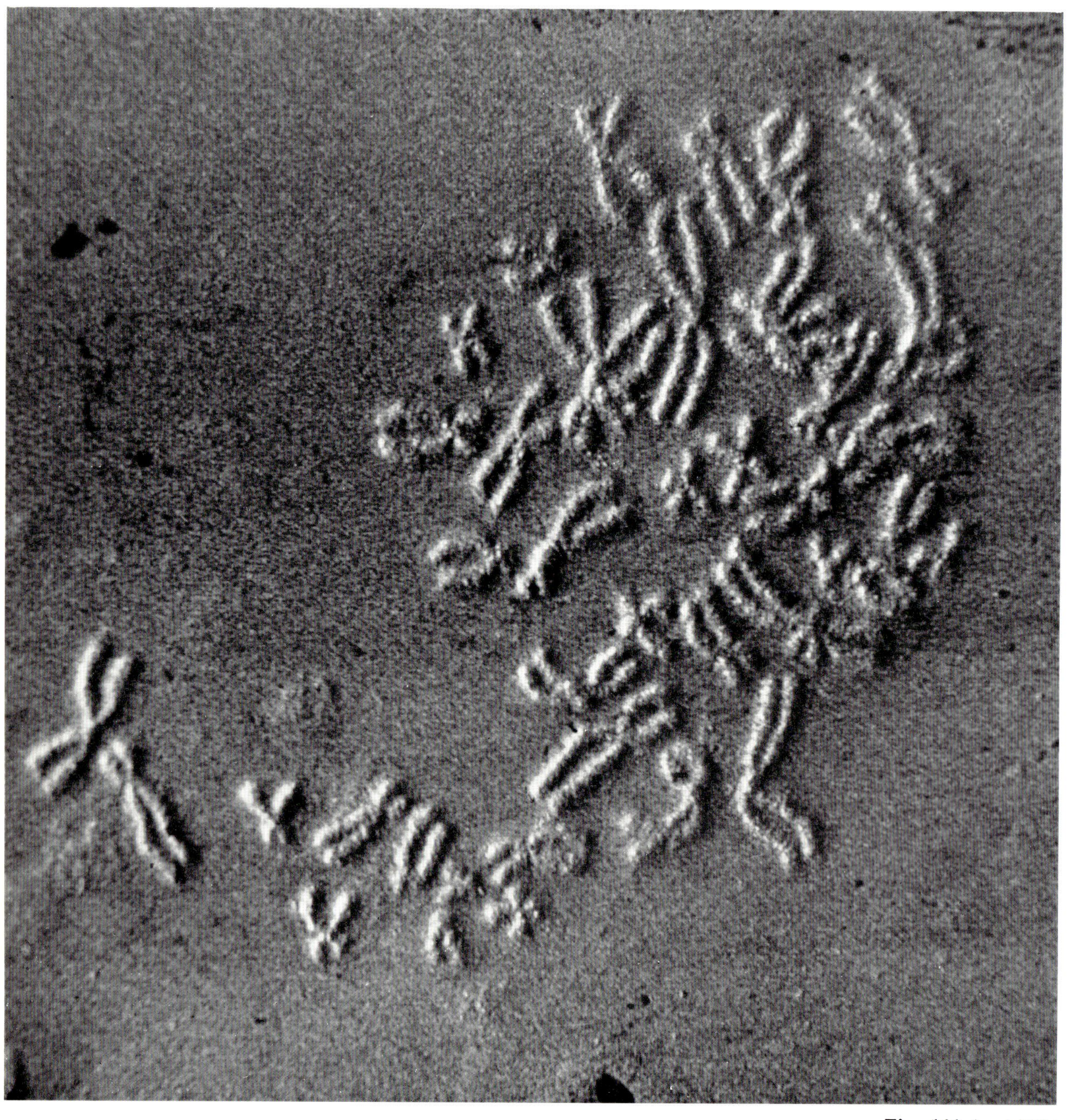

Fig. 144 ($\times$ 1,700)

Fig. **145** (×3,000)

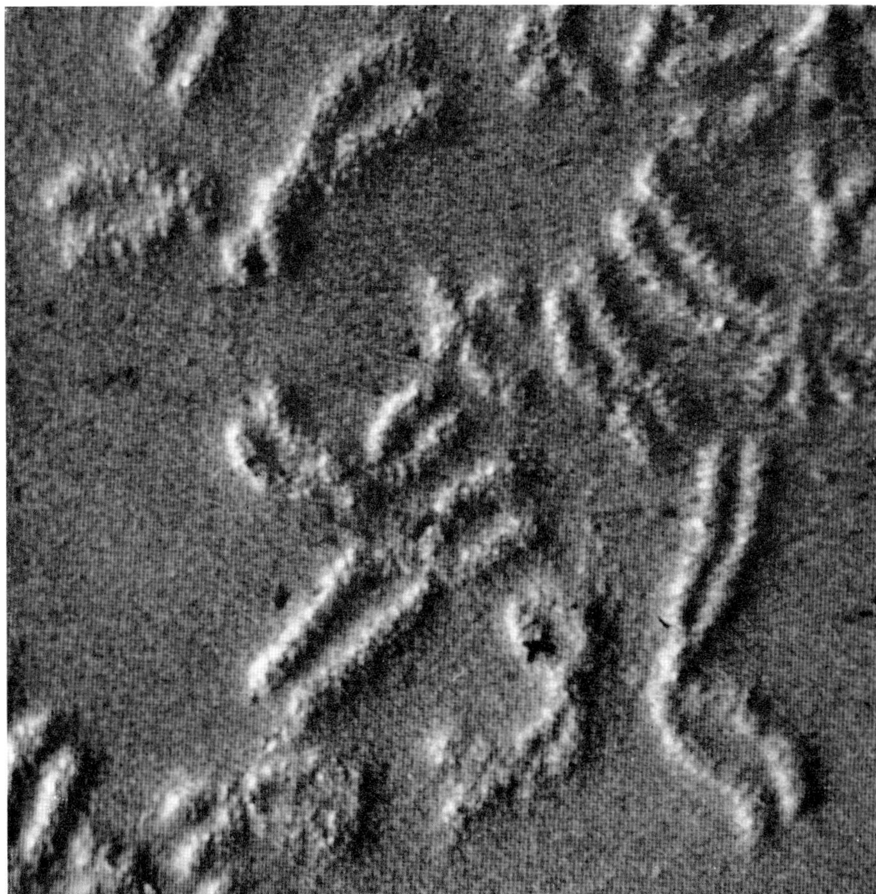

The micrographs and data of this section were provided by courtesy of Dr. Kéiichi Tanaka, Professor of Anatomy of the Tottori University Medical School.

Note

Circulating blood was cultured for 72 hours and treated with colchicine. Cells were swollen with 0.5 per cent sodium citrate, fixed in Carnoy and dropped on glass slides to be dried in air. Treatment in deoxycholic acid (5 mg/cc) for 2 hours and Giemsa staining were made for light microscopy prior to coating with gold. EM: JSM-2

Reference

Clarke, J.A., G.F. Rowland and A.J. Salsbury: The surface and internal structure of metaphase chromatin. Chromosoma 29: 74-87 (1970).

Neurath, P.W., M.G. Ampola and H.G. Vetter: Scanning electron microscopy of chromosomes. Lancet 7530: 1366-1367 (1967).

Pawlowitzki, I.H., R. Blaschke and R. Christenhusz: Darstellung von Chromosomen im Raster-Elektronen-mikroskop nach Enzymbehandlung. Naturwissenschaften 55: 63-64(1968).

Tanaka, K., R. Makino and A. Iino: The fine structure of human somatic chromosomes studied by scanning electron microscopy and the replica method. Arch. histol. jap. 32: 203-211 (1970).

Figs. 144-146 by courtesy of the Arch. histol. jap.

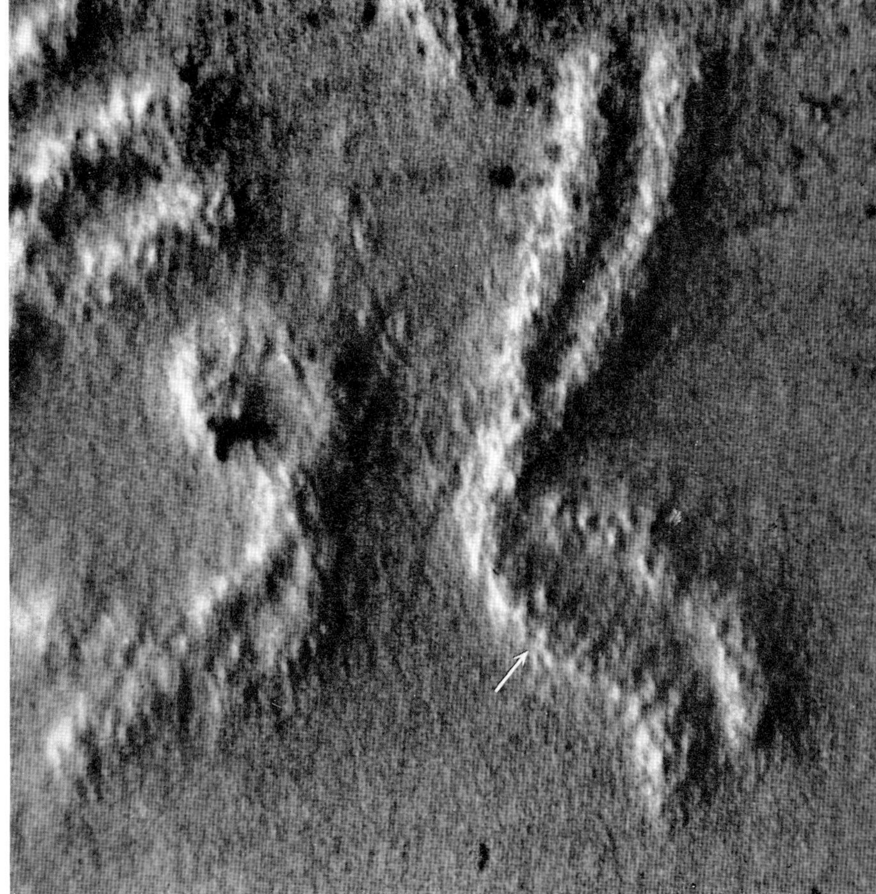

Fig. 146 (×6,400)

GALL AND URINARY STONES

35. Gall Stone

Gall stones cause one of the most common abdominal disturbances, one which is characterized by severe pains and inflammatory signs. Gall stones are divided by their chemical composition into two types. The cholesterol type is common in Caucasian races who eat much fat, while the bilirubin type is prevalent in Asian races who consume a smaller amount of fat. Recent studies indicate, however, that the majority of stones fall into a mixed type and their chemical content is generally complicated. Examination of gall stones with the scanning electron microscope is believed to be able to serve in filling a gap in the knowledge gained from light microscopic examination and from

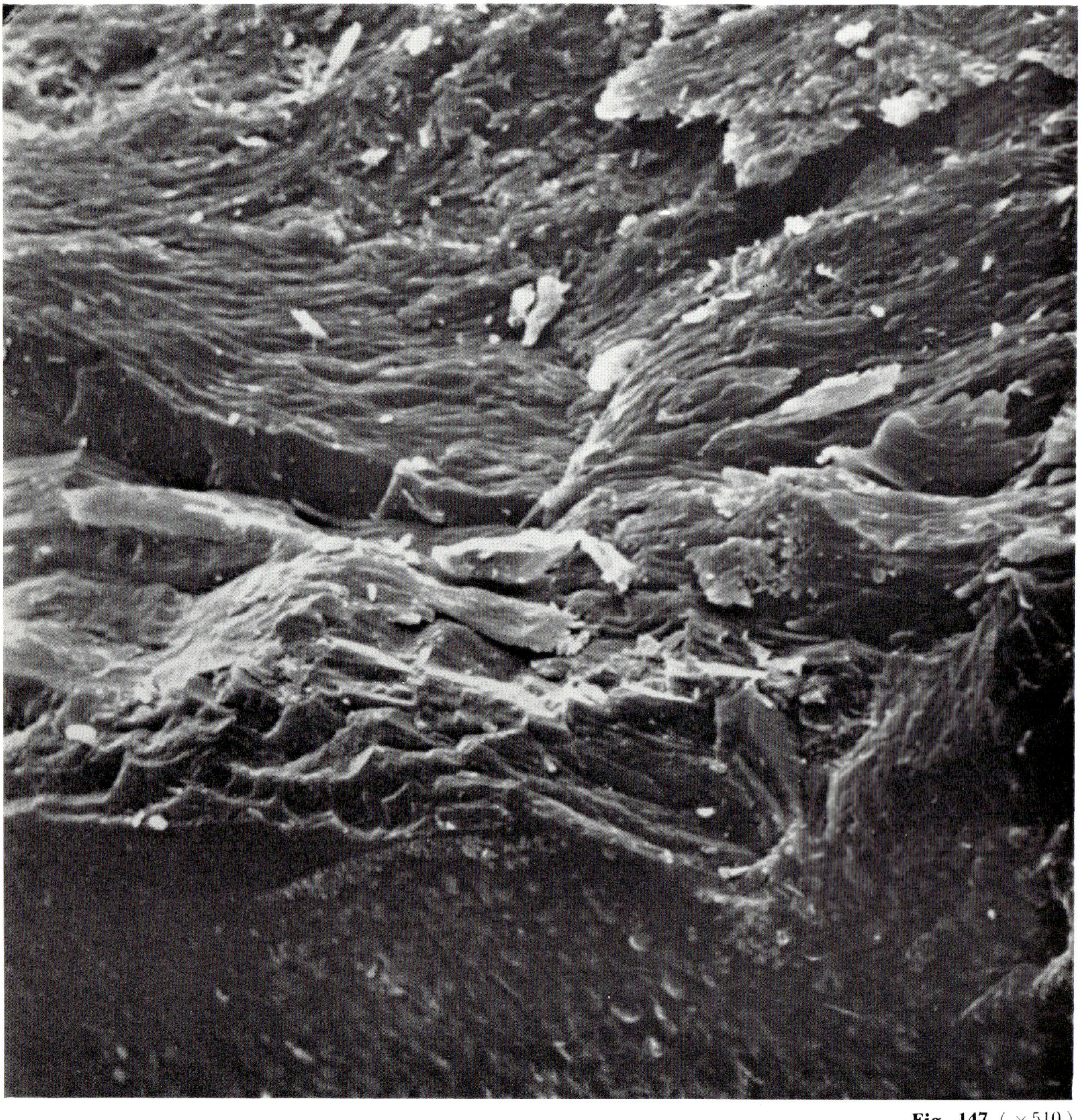

Fig. 147 (×510)

Fig. 148 (× 3,600)

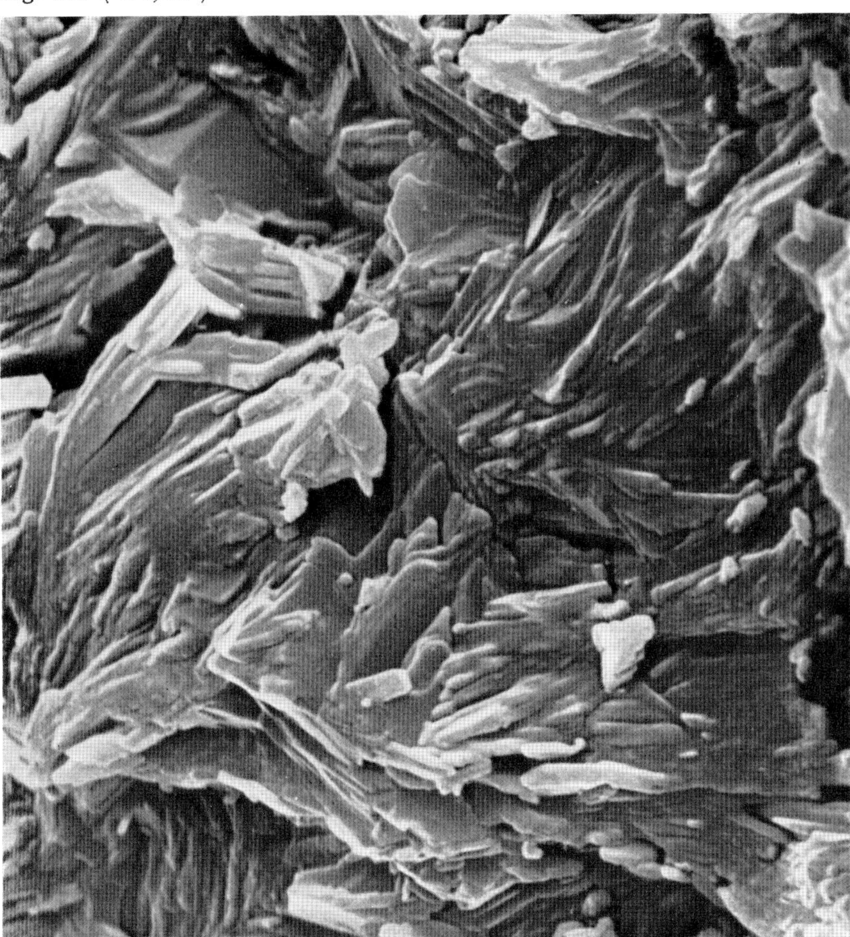

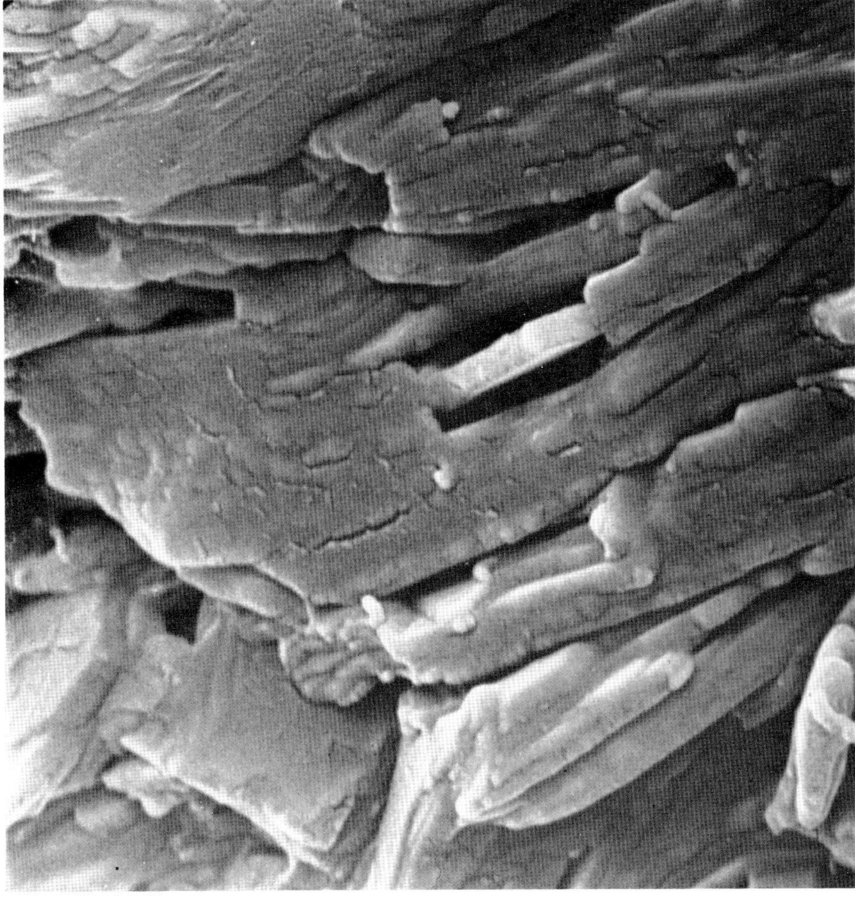

Fig. 149 (× 12,000)

chemical analysis of the stones.

The material shown here derives from a surgical operation on a 43-year-old woman. It was a mixed type stone appearing green-brown on the surface and pale yellow in the core.

Fig. 147 shows the broken surface of the cortical part. Slate-like crystals are piled up as in a twisted geological stratum. **Figs. 148** and **149** show the core part in closer views. Slate-like crystals are arranged in irregular directions and form a rather loose structure.

Bacteria, fungi, parasites and their eggs may be enclosed in gall stones. these forms which either survive in the stone or remain dead and coated with a film of bilirubin and other substances seem to be interesting objects to be studied with the scanning electron microscope.

Note

Broken surface of a gall stone was coated with carbon and gold. EM: JSM-2

Reference

OGATA, T. and F. MURATA: Scanning electron microscopic studies of the ultrastructure of human bile, pancreas and urinary stones (in Japanese). Rinsho Geka (Clinical Surgery) 25: 15-24 (1970).

36. Urinary Stone

That urinary stones have given human beings pain since very old times is known from the fact that they have been found in Egyptian mummies of 4000 BC. Their relation to medicine is also very old. The first pages of surgery are adorned by the endeavors and dramatic successes of surgeons in the fight against urinary stones. For modern surgery urolithiasis is one of the most simple diseases but our knowledge of the genesis and nature of urinary stones is still insufficient and a definite causal therapy is not known. Scanning electron microscopy is expected to contribute to the study of the natural history of stones, especially when combined with an X-ray microanalysis.

Fig. 150 (×510)

Fig. 151 (× 3,600)

Here under the scanning electron microscope is shown the broken surface of a stone which was taken in surgery from the ureter of a 40-year-old man. It measured 10 mm in diameter and was brown in color. In chemical examination this stone was found to be of a mixed type containing uric acid and its salts, oxalic acid and some calcium salts. (Chemical analysis was kindly made in the Department of Urology of the Okayama University Medical School.)

Fig. 150 indicates in a low-power micrograph that the stone consists of two zones of different structures. The central part or core which occupies the upper right of Fig. 150 is composed of slate-like crystals as shown at higher magnification in **Fig. 151.** The peripheral zone, on the other hand, is formed of prismatic crystals which are shown in a close-up in **Fig. 152.**

Note

Broken surface of a urinary stone was coated with carbon and gold. EM: JSM-2

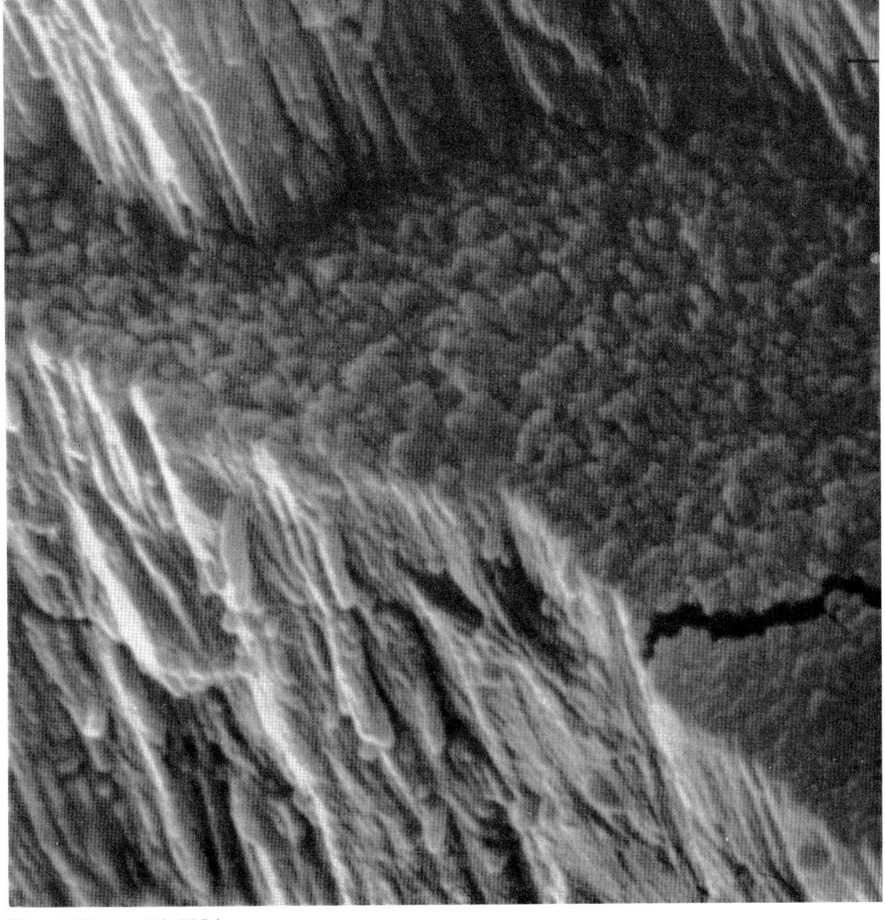

Fig. 152 (× 12,000)

DENTAL TISSUES

37. Enamel Formation in the Human Tooth

Enamel covers the crown of the tooth and represents the hardest substance found in the body. Enamel consists of bundles of rods which run radially through its whole thickness. Preceding the eruption of the tooth, enamel is generated by special cells called ameloblasts which form an epithelial hood on the bud of the tooth. The ameloblast is prismatic in form with a slightly thinned terminal portion called Tomes' process. The material of the enamel rods is produced at the end of this process much as dental cream is pushed out of a tube. The ameloblasts retreat as they deposit the rods. It is presumed that interprismatic enamel, a substance which fills the spaces among the rods is formed from the side surface of the Tomes' process.

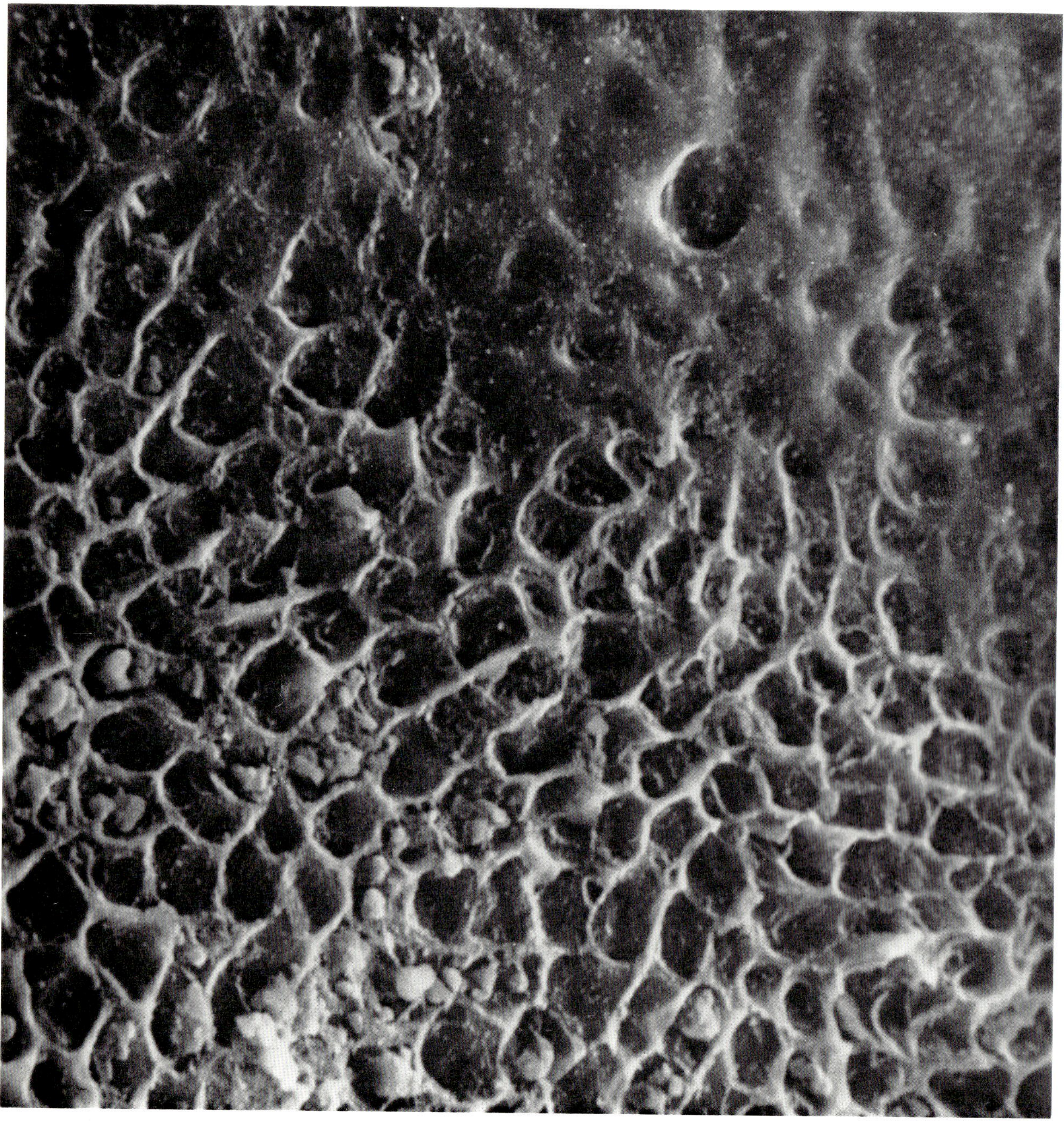

Fig. 153 (× 1,700)

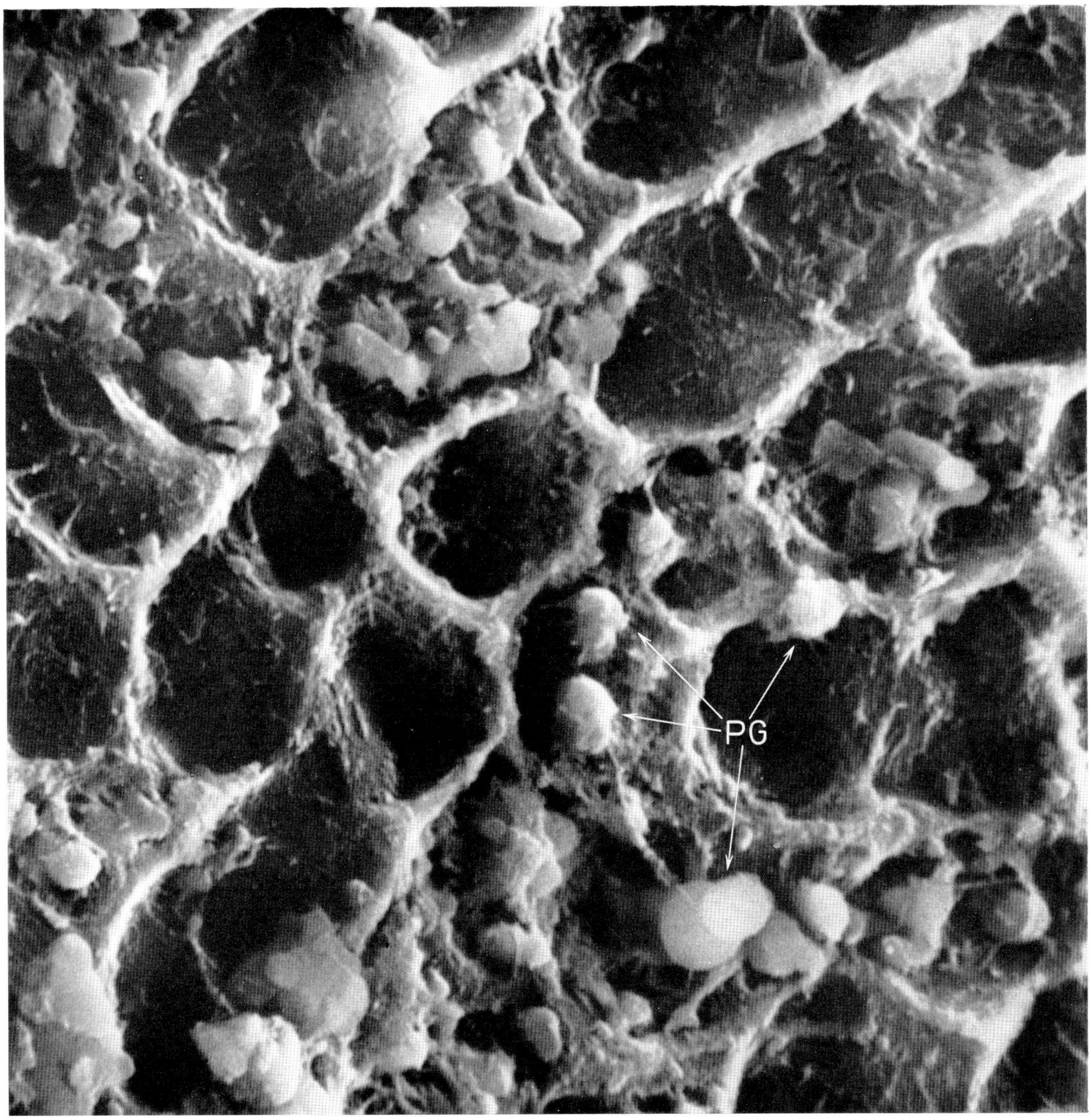

Fig. 154 (×5,100)

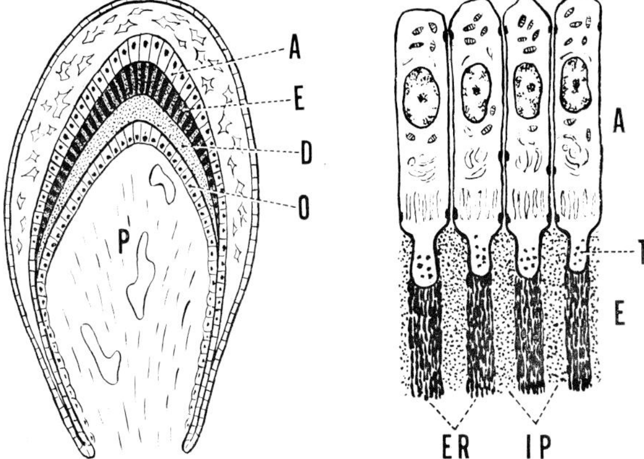

Fig. 155. Diagrams of a developing tooth (left) and the site of enamel formation (right).
A: ameloblasts forming enamel (**E**). **O:** odontoblasts forming dentin (**D**). **P:** primitive pulp. **T:** Tomes' process. **ER:** enamel rods. **IP:** interprismatic substance.

Fig. 153 is a scanning electron micrograph of the surface of the enamel of an erupting human tooth. In the major part of the picture, enamel formation is actively taking place, while in the relatively smooth portion in the upper right it seems to have ceased.

In the honeycomb-like relief of the enamel surface, each groove represents space which was filled with a Tomes' process in a living state and its base corresponds to the end of an enamel rod (**Fig. 154**).

The small, round bodies labeled "PG" are thought to be protein granules secreted by the ameloblasts. These granules may represent the base on which acicular crystals of apatite are deposited.

Fig. 156 (× 3,600)

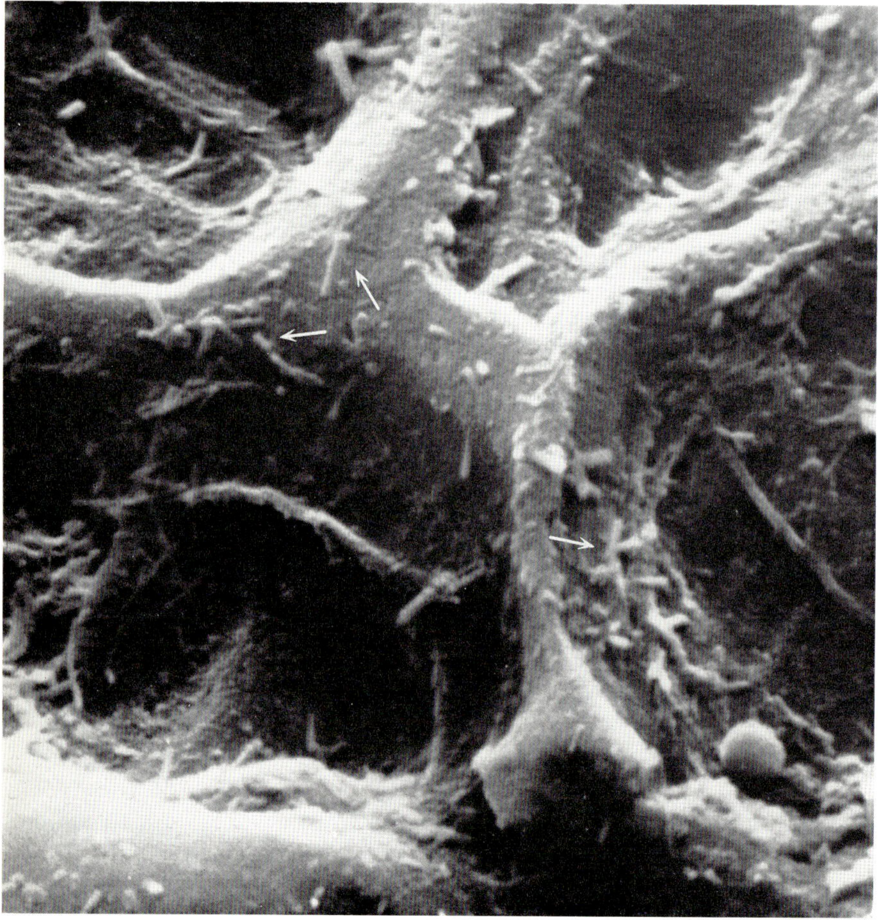

Fig. 157 (× 12,000)

In accordance with the shape of cross-sectioned rods characteristic of human enamel, the grooves appear mostly in the form of fish scales or of fans laid out with their arched side in a uniform direction. This pattern is especially clear in **Fig. 156.**

A closer view of the indentations is given in **Fig. 157.** One recognizes tiny rods (arrows) dispersed on the surface, both in the groove and on the ridge. They correspond to the apatite crystals, appearing as the initial forms of calcification in the deposit of enamel material.

Figs. 158 and **159** show the relatively smooth parts adjacent to the enamel under active formation.

Note

An incisor was taken from a dead fetus previously fixed in 10 per cent formalin. Soft tissues including the ameloblast layer were carefully removed with forceps. The enamel thus exposed was gently washed, dehydrated in acetone, dried in air and coated with carbon and gold. EM: JSM-2

Reference

BOYDE, A.: The structure of developing mammalian dental enamel. In: (ed. by) M. V. STACK and R.W. FEARNHEAD: Tooth enamel. Bristol, John Wright and Sons, 1965 (p. 163-167).

BOYDE, A.: The development of enamel structure. Proc. Roy. Soc. Med. 60: 923-928 (1967).

BOYDE, A.: Correlation of ameloblast size with enamel prism pattern: Use of scanning electron microscope to make surface area measurements. Z. Zellforsch. 93: 583-593 (1969).

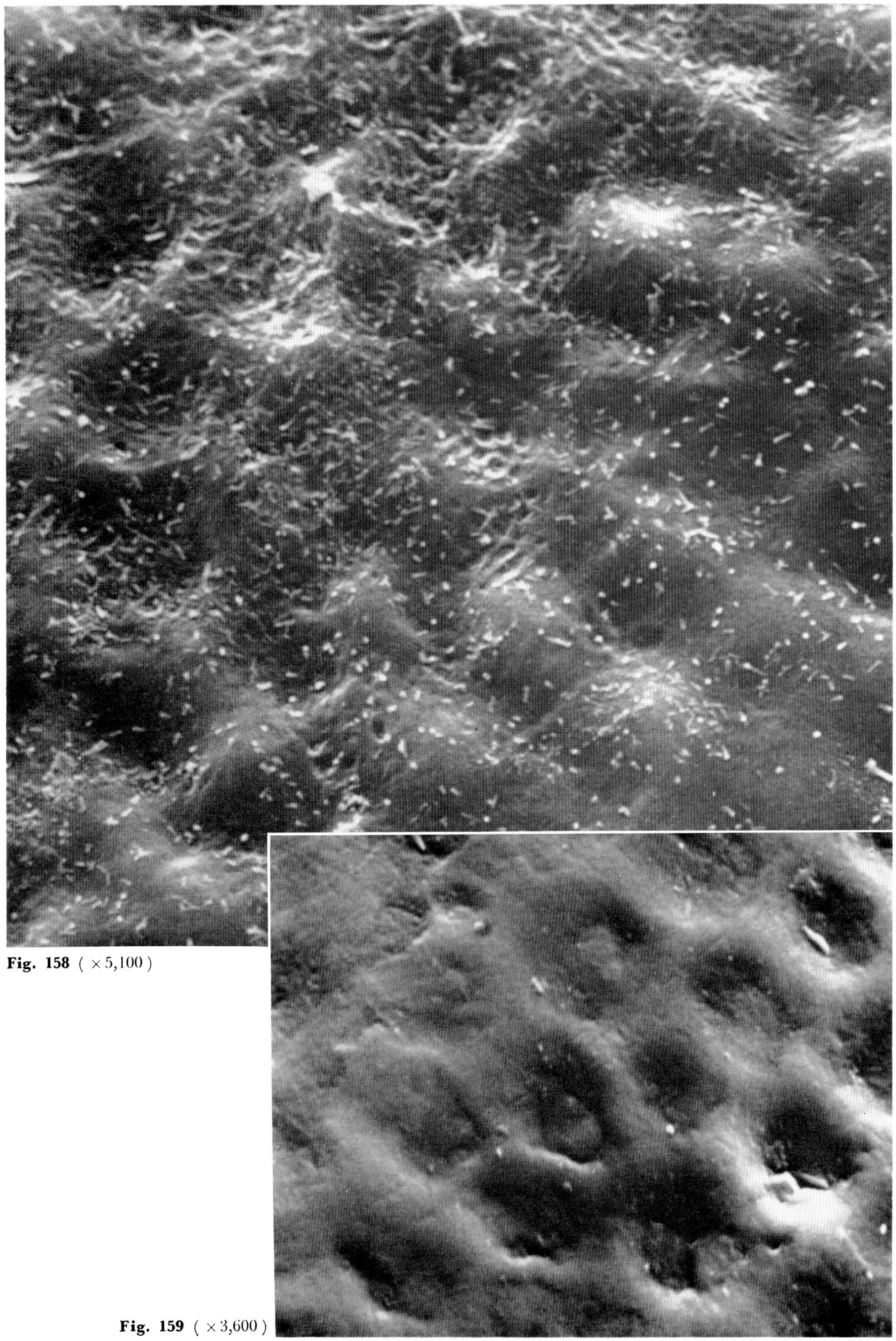

Fig. 158 (×5,100)

Fig. 159 (×3,600)

38. Enamel Surface of the Human Tooth

The surface of the tooth is not smooth when observed under the microscope. There remain those grooves corresponding to the ends of individual enamel rods (see p. 106) though much less conspicuous than in the enamel under formation. A gentle undulation is formed of periodic elevations and furrows running horizontally. This undulation, called perikymata, is known to be caused by the meeting of the enamel surface and the Retzius' stripes which are the growth lincs or, so to speak, the "annual rings" of the enamel. (The Retzius' stripes are formed not one in every year as annual rings, but one in every ten days.)

Fig. 160 shows the buccal surface of a third lower molar. As this tooth was still covered by the mucous membrane, the enamel surface is free of the wearing effect of mastication. The grooves are clear only in the furrows of the undulation. A sharp transverse line across the undulation likely represents an enamel lamella which is generally thought to be a fissure filled with organic substances.

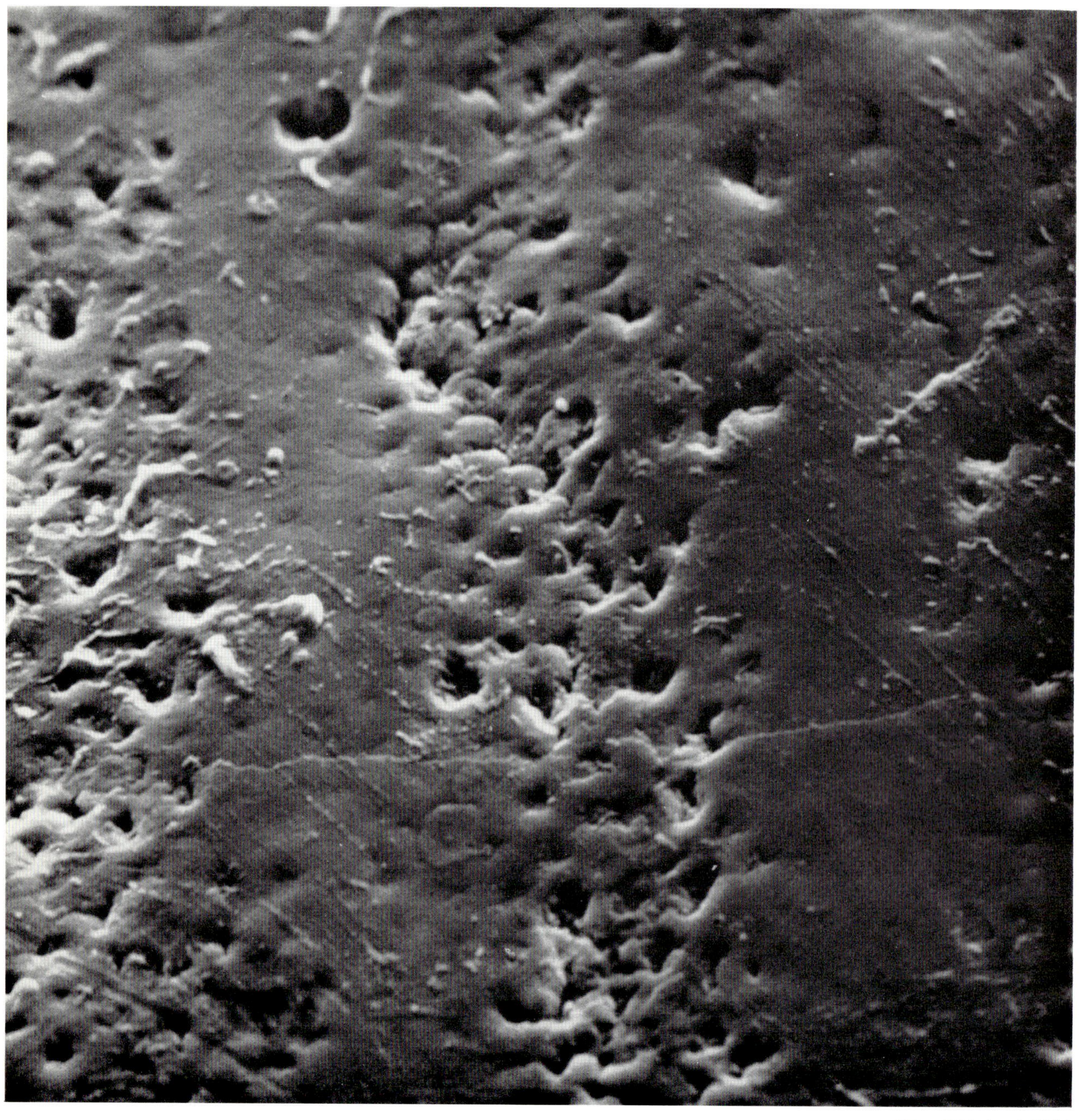

Fig. 160 (× 1,700)

Fig. **161** ($\times 3{,}600$)

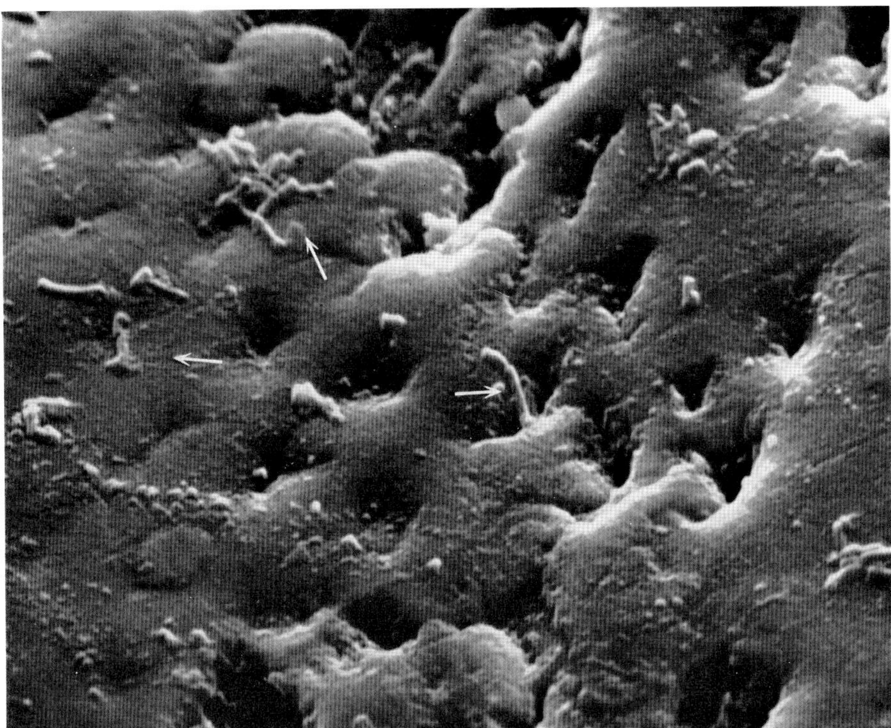

Fig. 161 is a closer view of the central area of Fig. 160. Worthy of notice are numerous rod-like bacilli attached to the enamel surface (arrows).

Fig. 162 is the enamel surface of a third molar slightly worn down by mastication. The undulation is clear, showing shallow but sharp furrows which may be called imbrication lines. Slight indentations indicate the ends of individual enamel rods.

Note

Carefully extracted teeth were fixed in 10 per cent formalin, dehydrated in acetone, dried in air and coated with carbon and gold. EM: JSM-2

Reference

BOYDE, A. and K. S. LESTER: The structure and development of marsupial enamel tubules. Z. Zellforsch. 82: 558-576 (1967).

HOFFMAN, S., W. S. MCEWAN and C.M. DREW: Demineralization studies of prefluoridated dental enamel surfaces by scanning electron microscopy. Proc. 2nd Annu. Scan. Electron Microsc. Symp. 1969 (p. 135-147).

VAHL, J. VON and G. PFEFFERKORN: Elektronenoptische Untersuchungen der durch Laser-Beschuß hervorgerufenen Veränderungen an Zahnhartsubstanzen. Deut. zahnärztl. Z. 22: 386-394 (1967).

VAHL, J. VON and H. RIEDEL: Mikromorphologische Strukturanomalien des Zahnschmelzes bei Mindermineralisation (Schmelzhypoplasie). Deut. zahnärztl. Z. 23: 289-297 (1968).

ZUNIGA, M. A. and M. B. QUIGLEY: Incipient carious enamel lesions studied by scanning electron microscopy. Proc. 3rd Annu. Scan. Electron Microsc. Symp. 1970 (p. 209-216).

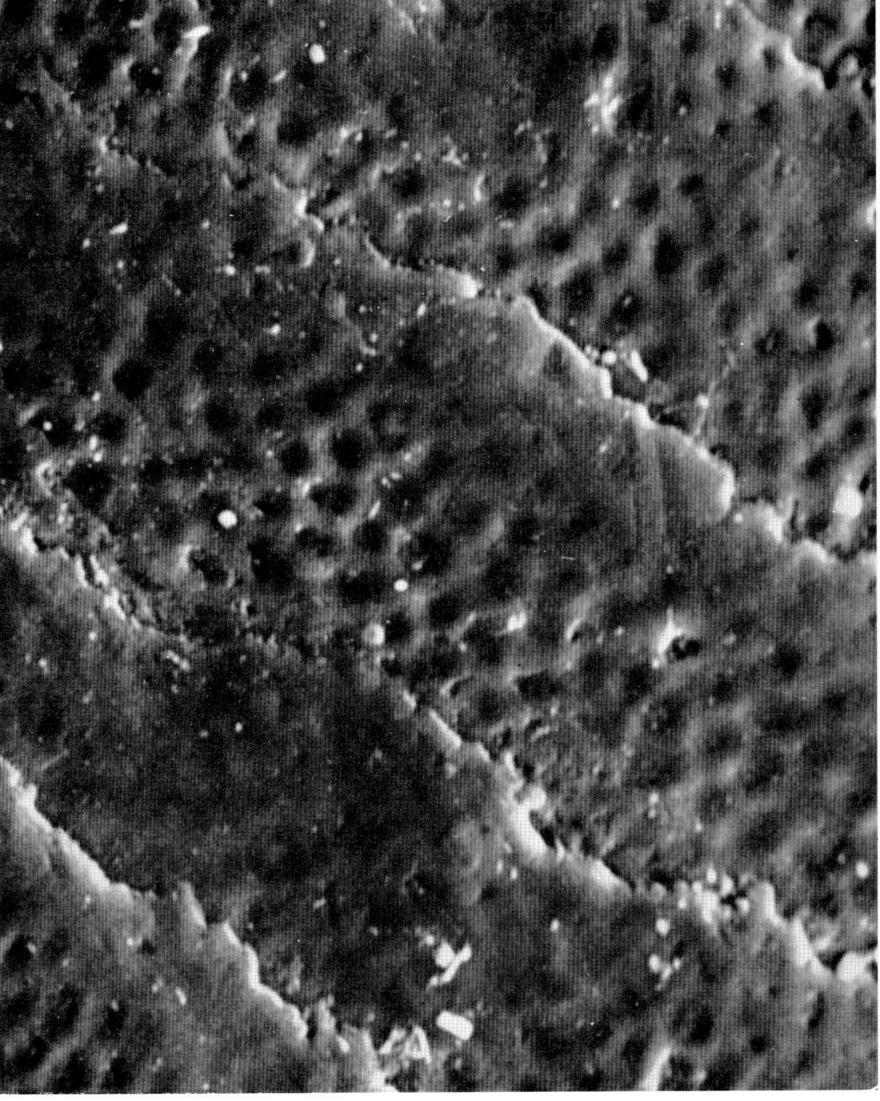

Fig. 162 ($\times 870$)

39. Surface of Enamel and Dentin Sections

The structure of tooth enamel and dentin can be demonstrated with the scanning electron microscope using the cut surface of the tooth, ground and slightly etched.

Fig. 163 shows enamel (left) and dentin (right) in a longitudinally cut human tooth. The cleft at the site of the dentino-enamel junction has occurred in the drying of the specimen.

In the enamel, bundles of enamel rods stand on the dentino-enamel junction like flames. It should be noticed, however, that the elevated structures in the micrograph are not the enamel rods themselves, but their sheaths or interprismatic substance which is more resistant to acidic etching than the rods. The rods appear as dark furrows in this picture. Because of the partly twisted course of the enamel rods and their sheaths, bundles of longitudinally hit structures and obliquely cut ones appear alternately along the dentino-enamel junction. This corresponds to the bands of Hunter-Schreger, the alternate light and dark zones in the longitudinal sections of the tooth under the light microscope.

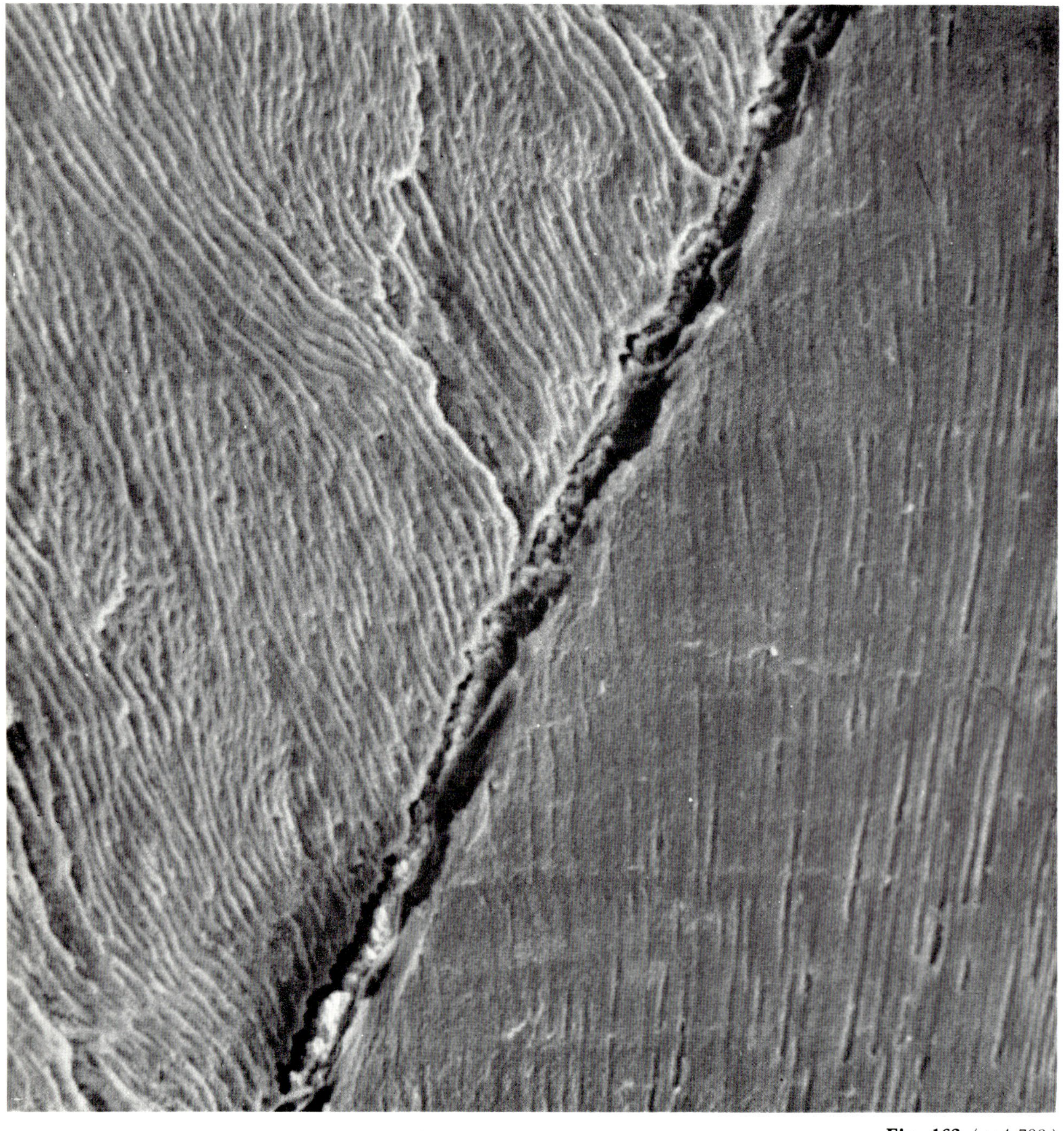

Fig. 163 (× 1,700)

Fig. **164** (× 3,600)

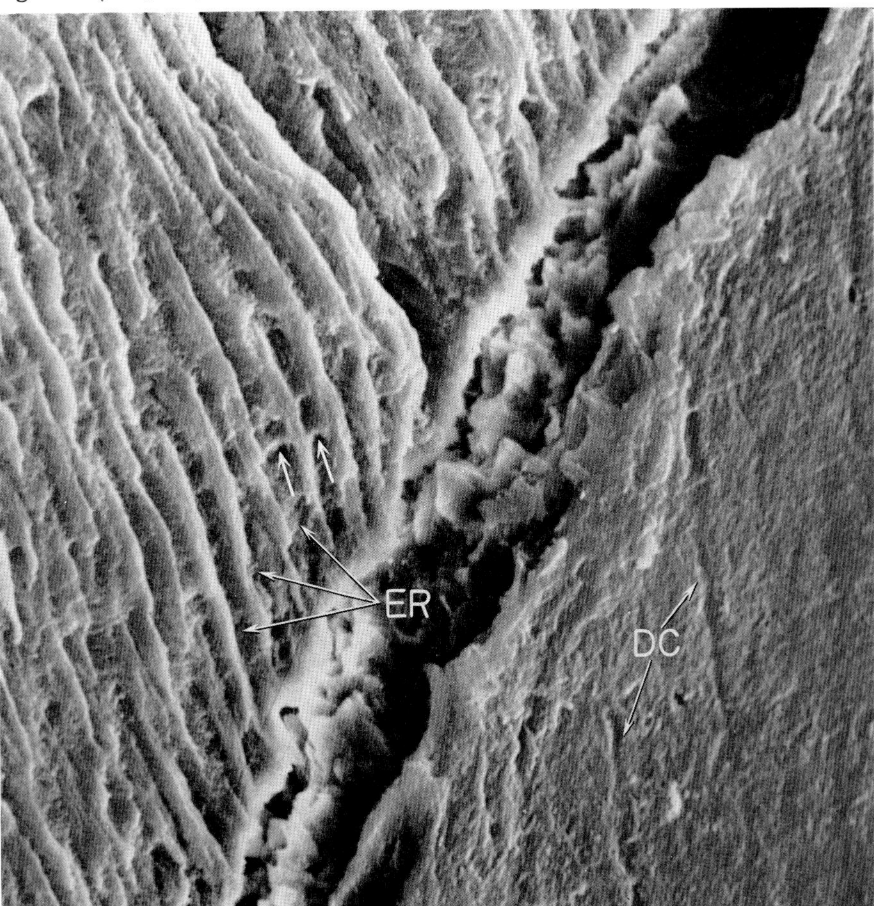

In the dentin, thin canals are hit longitudinally or obliquely. They are called dentin canalicules and contain, in a living state, fibers of Tomes which are the long processes of the odontoblasts, the cells which generate the dentin tissue.

Fig. 164 shows the central part of Fig. 163 more closely. Enamel substance ensheathing individual rods (interprismatic enamel) is seen as elevated columns, whereas the rods (ER) themselves have sunk into furrows by etching. Striations of the rods are recognized as shelves across the furrows (arrows). Dentin canalicules are labeled "DC".

Fig. 165 shows a cut surface of enamel below the broken line and the free surface above the line. Both in cut and free surfaces the indented portions correspond to enamel rods.

Note

The cut surface of an extracted lower central incisor of an adult was ground and etched with 0.1 per cent hydrochloric acid for a few seconds. The specimen was dehydrated in acetone, dried in air and coated with carbon and gold. EM: JSM-2

Reference

BOYDE, A.: Electron microscopic observations relating to the nature and development of prism decussation in mammalian dental enamel. Bull. Group. Int. Rech. Sc. Stomat. 12: 151-207 (1969).

BOYDE, A. and S. J. JONES: Scanning electron microscopy of cementum and Sharpey fibre bone. Z. Zellforsch. 92: 536-548 (1968).

FROMME, H. G. VON, H. RIEDEL and J. VAHL: Beobachtungen zum Gefügeaufbau der Zahnhartsubstanzen mit Hilfe des Raster-Elektronen-Auflichtmikroskopes. Deut. zahnärztl. Z. 22: 395-400 (1967).

LESTER, K. S. and A. BOYDE: The surface morphology of some crystalline components of dentin. In: (ed. by) N. B. B. SYMONS: Dentine and pulp; their structure and reactions. Edinburgh, Livingston, 1968.

LESTER, K. S. and A. BOYDE: Some preliminary observations on caries ("remineralization") crystals in enamel and dentin. Virchows Arch. Abt. A, Pathol. Anat. 344: 196-212 (1968).

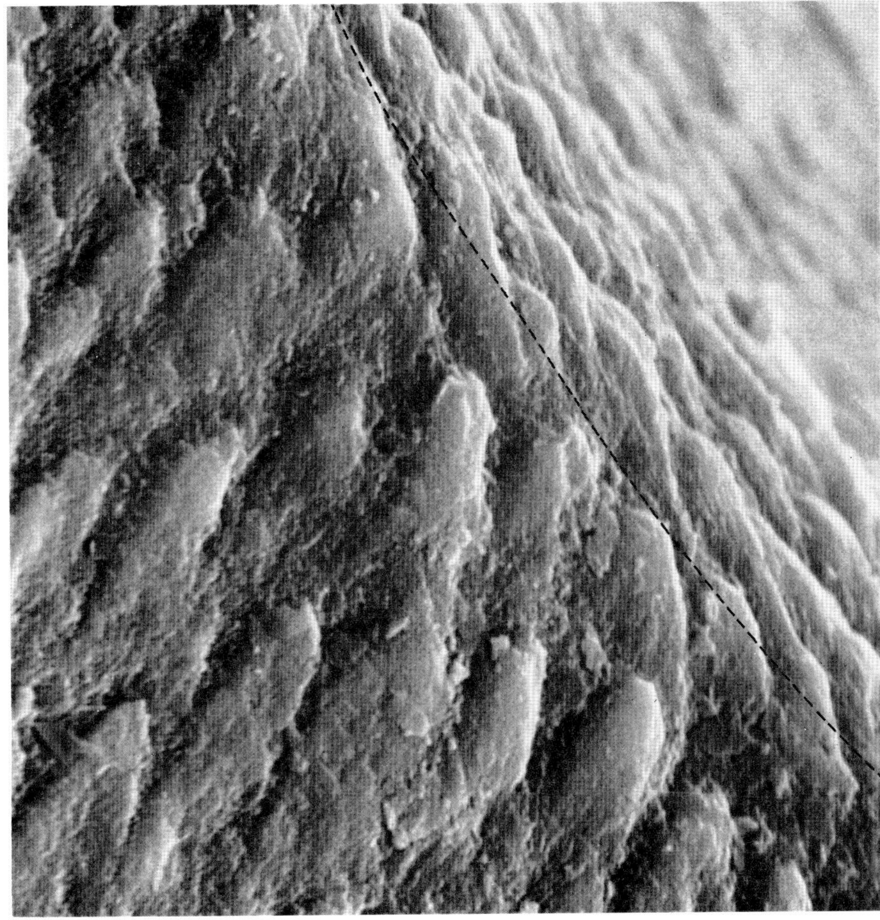

Fig. 165 (× 12,000)

40. Secondary Dentin

As the designation indicates, dentin comprises the main and basal part of the tooth (dens: tooth). As compared with the enamel which is rather a dead mineral substance, dentin is a living matter penetrated by the Tomes' fibers, which are long processes of cells called odontoblasts. When a local erosion in caries or defacement by an excessive use of toothbrush reaches, after scooping out the layer of enamel, the outer surface of dentin at a certain site, odontoblasts lying on the opposite or pulpar surface of the dentin perceive this accident by an irritation on their processes and reactively form a new dentin mass, secondary dentin, on the corresponding site on the pulpar surface. Thank to this compensatory formation of secondary dentin, the tooth is not easily worn down to the pulp as the erosion gradually proceeds.

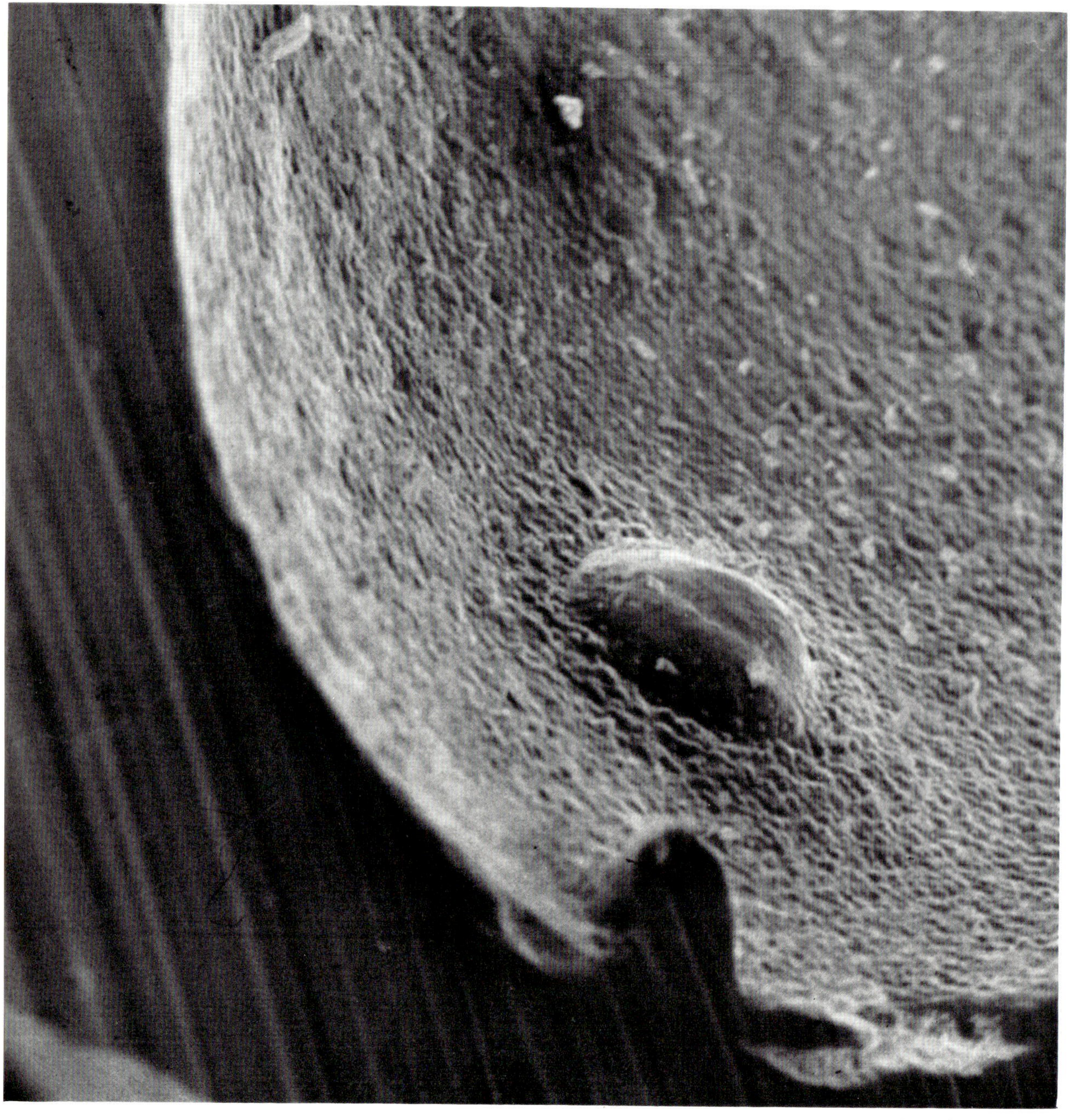

Fig. 166 (× 170)

Fig. **167** (×360)

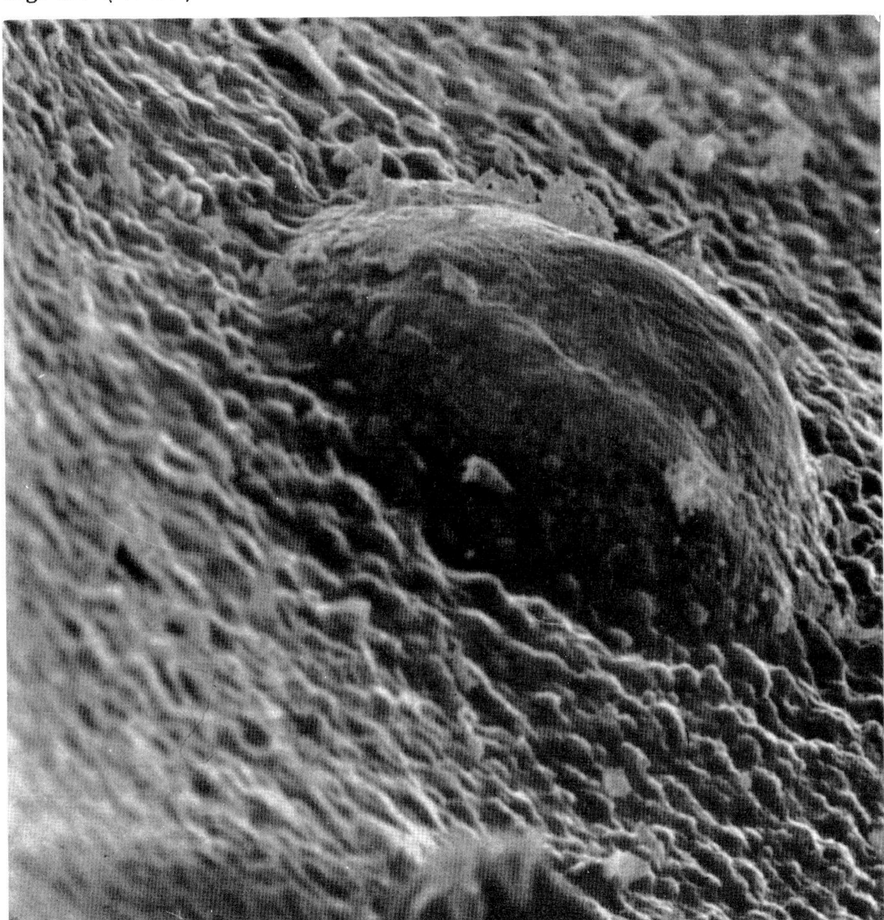

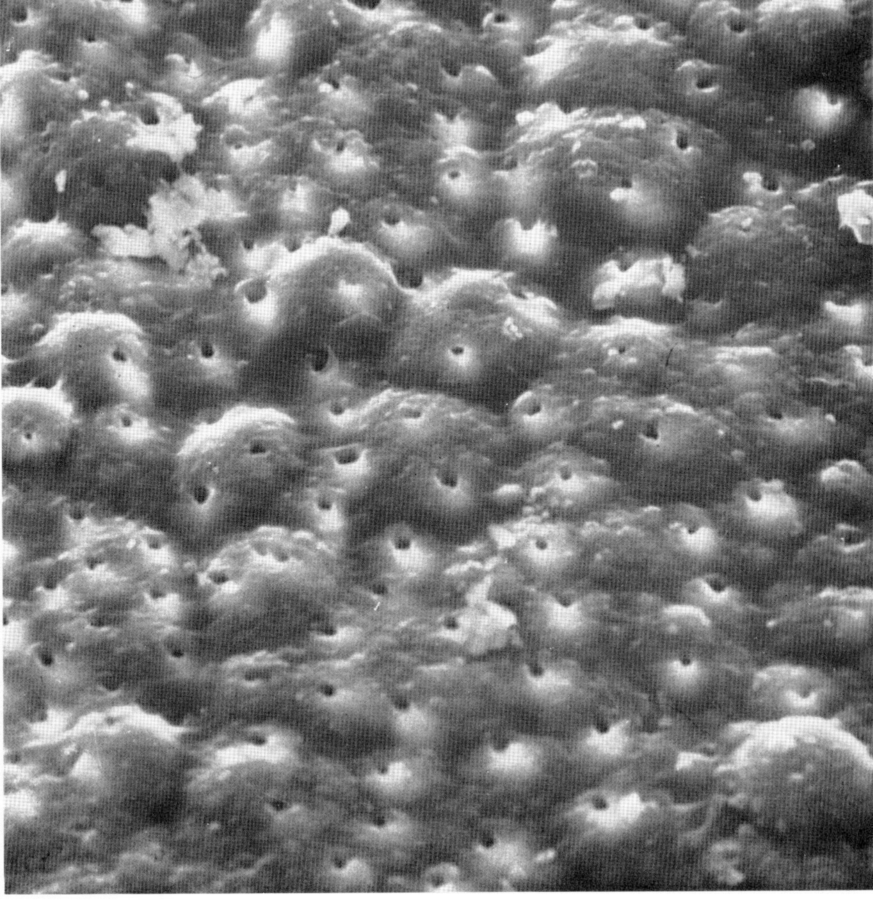

Fig. 168 (×1,200)

Fig. 166 shows the interior wall of dentin in an incisor cut longitudinally and cleansed of the pulp. The whole pulpar surface of dentin is covered with fine elevations. A large, lumpy elevation which is more closely shown in **Fig. 167** is the accumulation of secondary dentin. Another elevation, nipple-shaped, has been just hit by sectioning but appears out of focus in this micrograph. These round-shaped dentin accumulations are called denticles.

Fig. 168 is a closer view of the pulpar surface of the ordinary or primary dentin. Small round elevations cover the surface and tiny pores on them indicate the orifices of dentin canaliculi which, in a living state, contained the Tomes' fibers of odontoblasts.

Note

An incisor extracted and fixed in 10 per cent formalin was cut longitudinally. After the pulp was carefully removed with forceps and a jet stream of distilled water, the specimen was dehydrated in acetone, dried in air and coated with carbon and gold. EM: JSM-2

41. Cement Surface and Dental Calculus

The dental cement forms a thin layer covering the root of the tooth and, together with a connective tissue layer called periodontium, intermediates the junction of the tooth (dentin) and the bone of the jaw. Simply it may be regarded as identical with bone tissue, but its microscopic structure is markedly variable according to its developmental condition, its site in the tooth and the functional state of the tooth concerned.

Fig. 169 is a low-power scanning electron micrograph of the cement surface between the forked roots of an impacted premolar. The whole surface, which is attached in a natural state to the periodontium, appears rough, covered with tiny grooves of about 1μ in diameter. Large craters (arrows) about 10μ in diameter are the lacunes in which individual cementocytes lived.

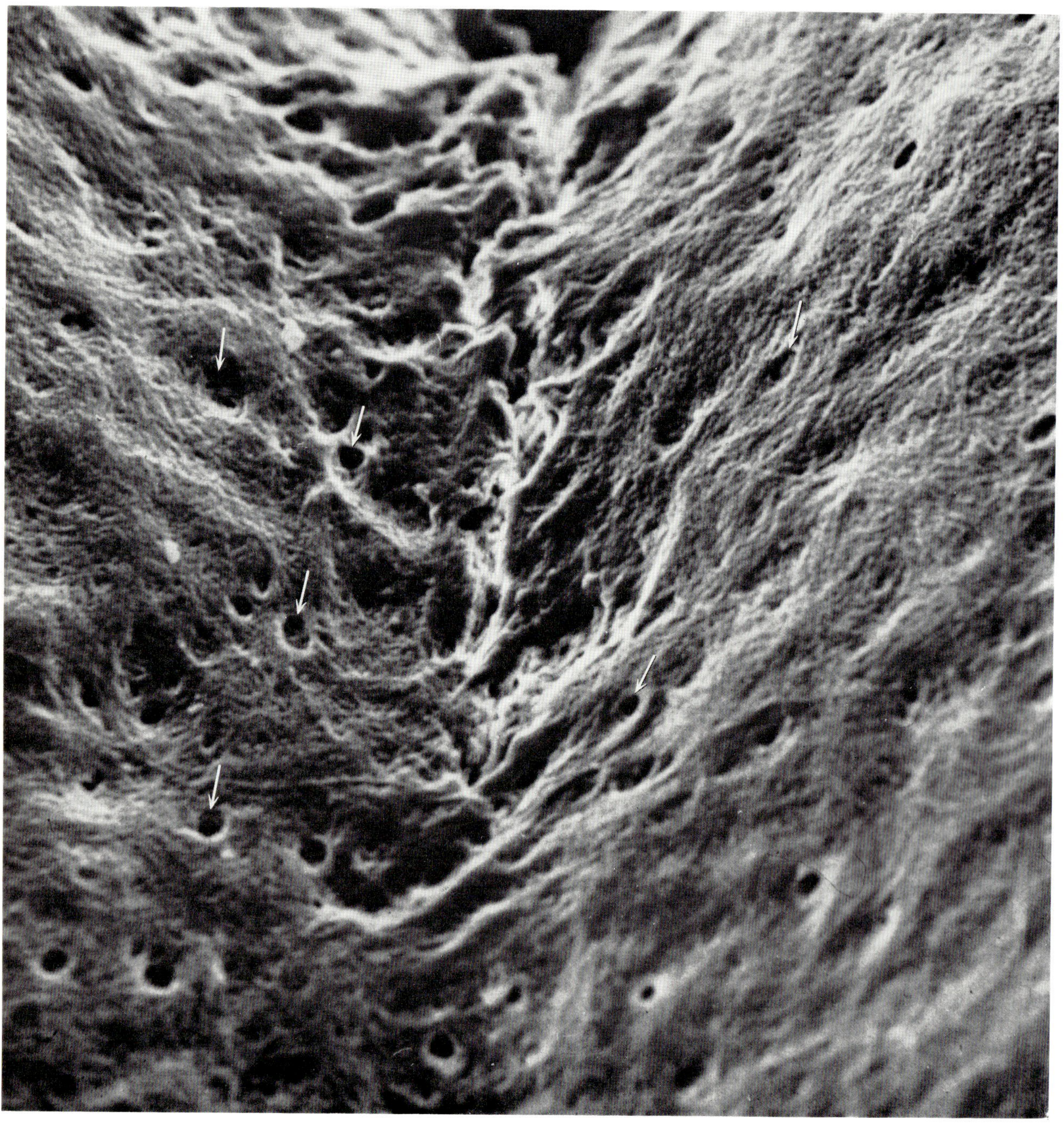

Fig. 169 (×510)

Fig. 170 (× 1,200)

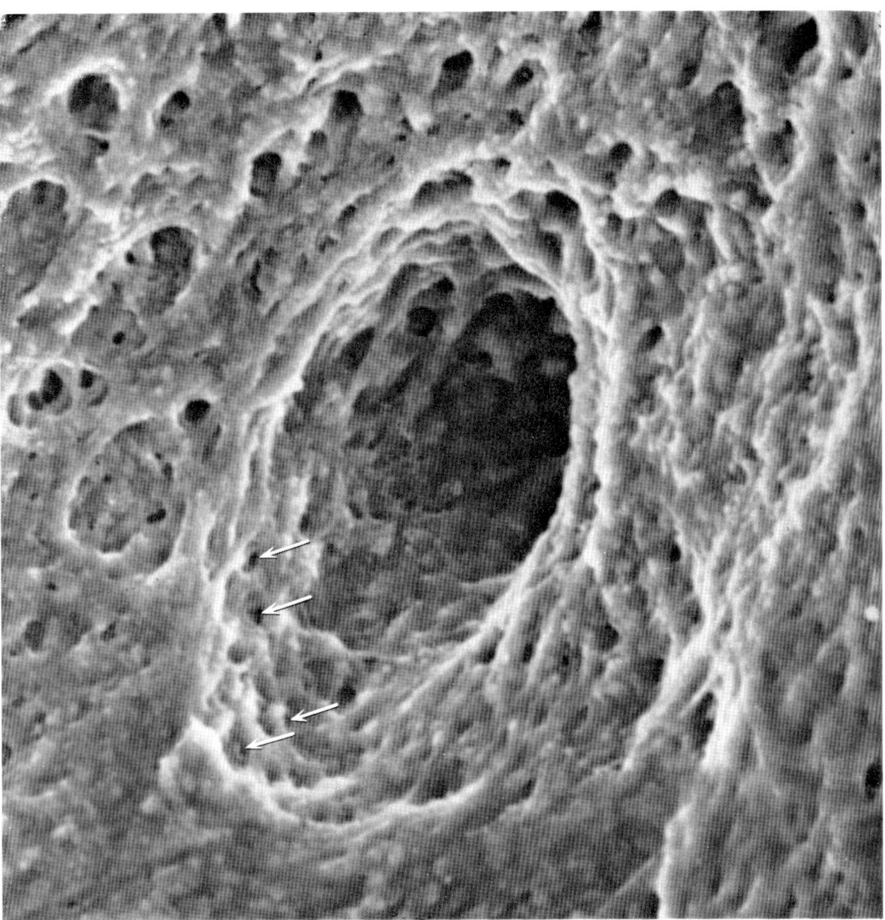

Thick periodontal fibers called Sharpey's fibers which bore into the cement, forming large grooves on its surface, are abundant in an erupted tooth but are not developed in this impacted tooth.

Fig. 170 is from a tooth in which a marked new-formation of cement was recognized. The large oval groove is believed to be the lacune of a cementocyte which was involved in the formation of the cement. Small pores whose orifices are margined with a ring-like structure (arrows) are thought to have received the fine processes of the cementocyte. The pores dispersed around the lacune likely contained Sharpey's fibers in the natural state.

Fig. 171 shows the boundary of cement (labeled "C") and enamel (E) in a second premolar from an 11-year-old boy. The surface of the cement which, before the tooth extraction, was exposed 2 mm above the gum shows winding and branching strings. This characteristic figure corresponds to dental calculus formed by calcification on the network of dead colonies of microbes such as actinomyces and leptotrichia.

Note

The tooth shown in Fig. 169 was extracted carefully and soft tissues were removed in warm, running water and then boiled for 30 minutes. The tooth in Fig. 170 was treated, after extraction, in 0.08M NaOH (100°C, 5 minutes). The tooth in Fig. 171 was fixed in 10 per cent formalin and boiled in water for 30 minutes. The specimens were dehydrated in acetone, dried in air and coated with carbon and gold. EM: JSM-2

Reference

BOYDE, A. and K. S. LESTER: Electron microscopy of resorbing surfaces of dental hard tissue. Z. Zellforsch. 83: 538-548 (1967).

TOKUNAGA, J. and T. MORI: A scanning electron microscopic observation of human cementum. Jap. J. oral Biol. 11: 13-23 (1969).

VAHL, J. VON, G. PFEFFERKORN and H.J. HÖHLING: Sublichtmikroskopische Untersuchungen am menschlichen Speichelstein. Deut. zahnärztl. Z. 23: 39-44 (1968).

Figs. 169-170 by courtesy of the Jap. J. Oral Biol.

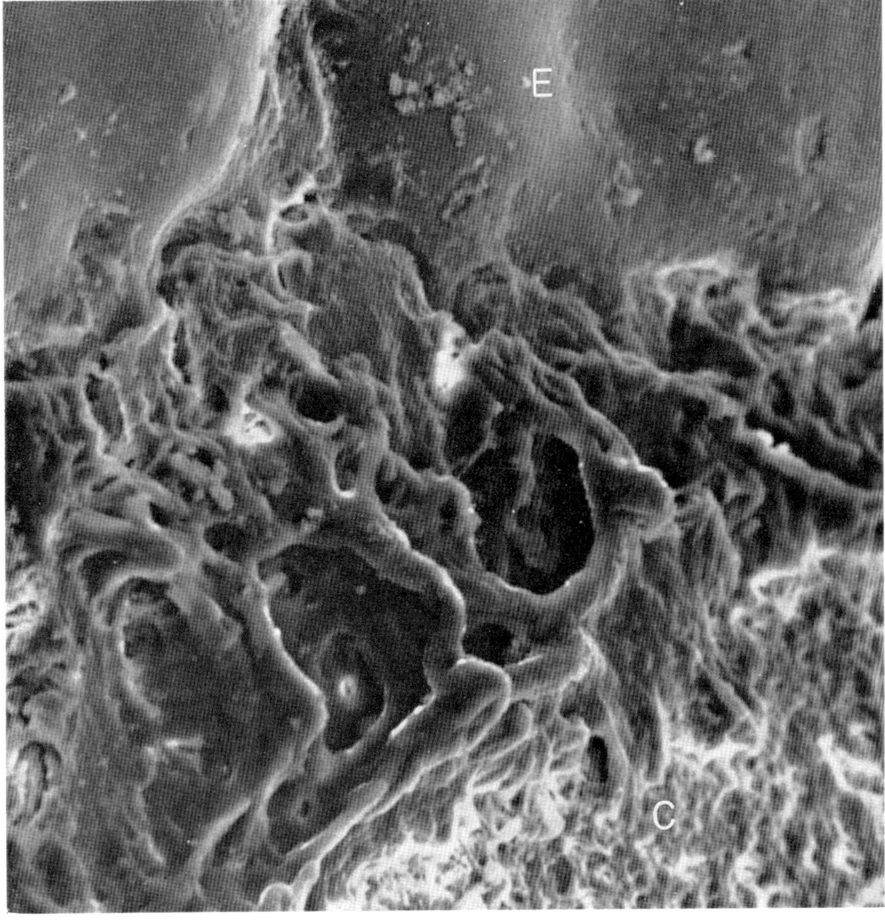

Fig. 171 (× 360)

PARASITES

42. Cercaria of Paragonimus

Paragonimus is a genus of trematode worms and is parasitic on the human lung, causing pulmonary distomiasis. Trematode worms in their final free-swimming larval stage are called cercariae. They invade the bodies of shell-fish, fish and crabs and develop to the infesting stage of trematodes for man.

A ventral view of a fully developed cercaria of *Paragonimus sadoensis* under the scanning electron microscope is shown in **Fig. 172.** A stylet (ST) and a tail (T) are seen on opposite ends of the body. Besides an oral sucker (OS), there is an acetabulum or abdominal sucker (A) which looks like a second mouth, and this is the origin of the old designation of this group of animal, "distoma" (two mouths).

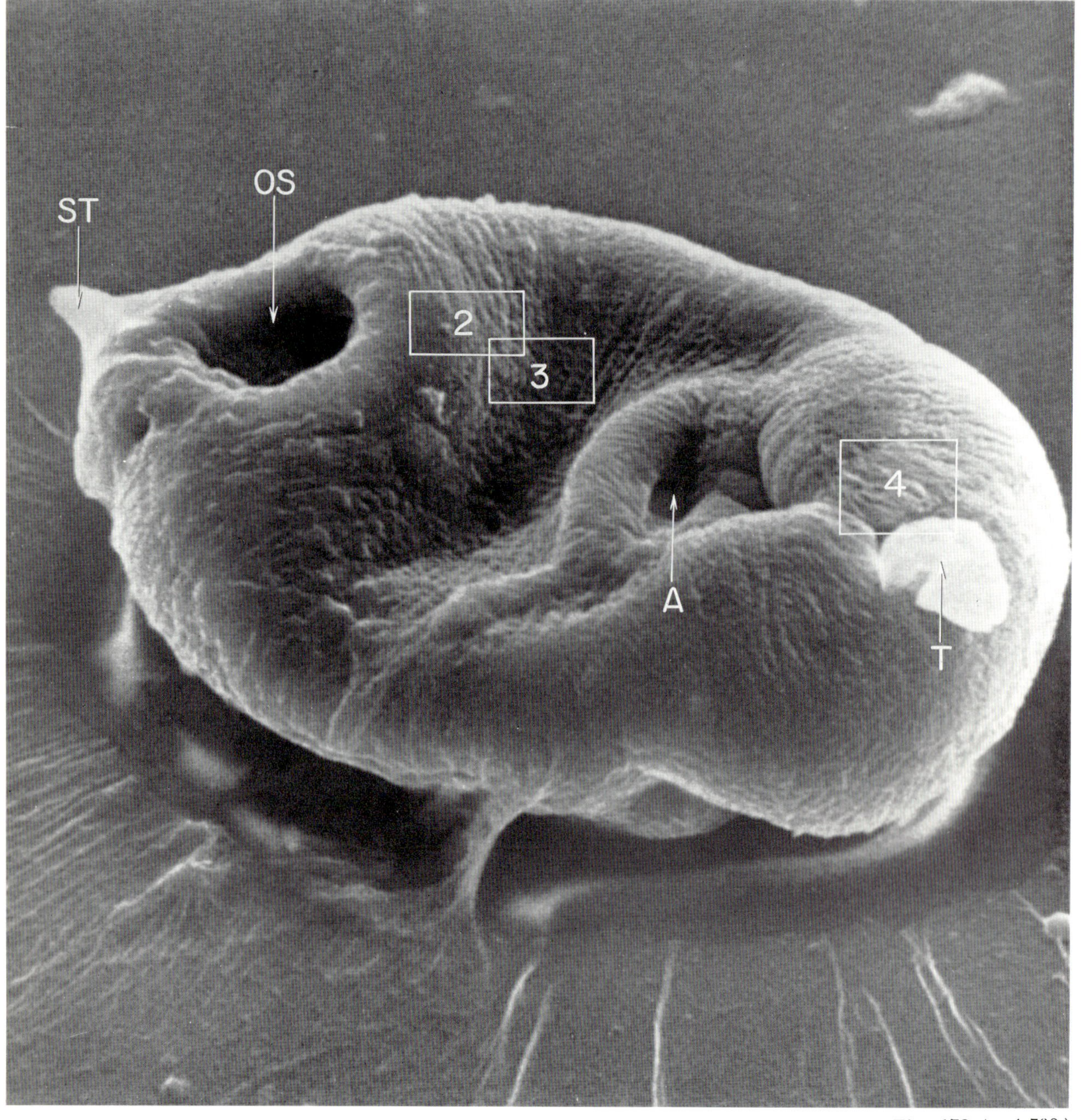

Fig. 172 (× 1,700)

Fig. **173** (× 12,000)

Fig. **174** (mid.) (× 12,000)

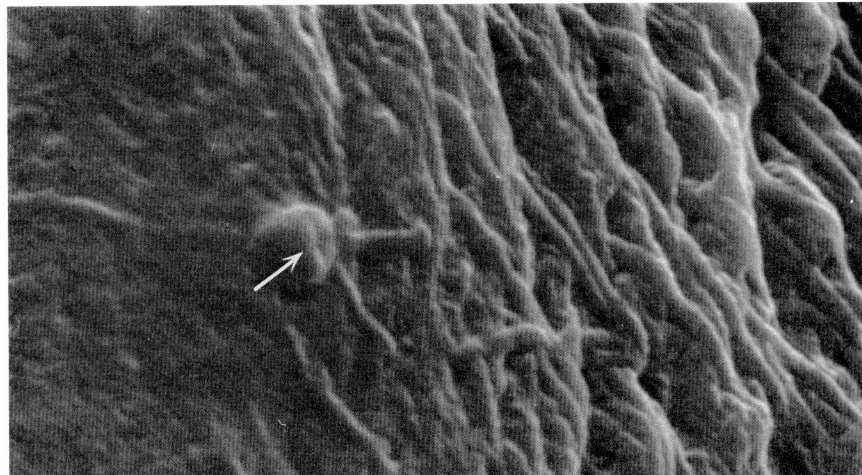

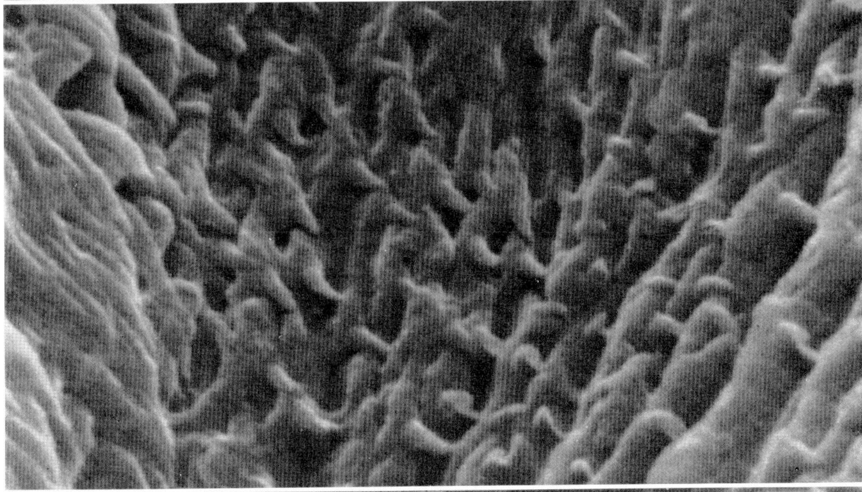

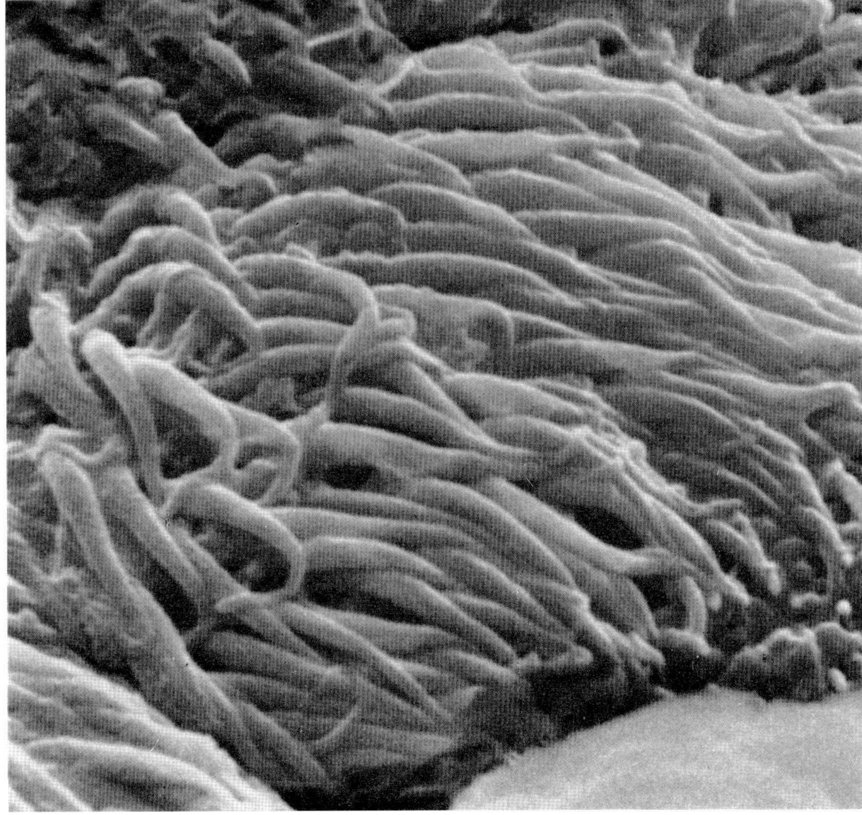

Fig. 175 (× 12,000)

A region behind the oral sucker (box "2") is shown in **Fig. 173** at higher magnification. The arrow indicates a sensory papilla. **Fig. 174** shows spines and cilia-like hairs in a region (box "3") between the two suckers. **Fig. 175** reveals the tapered hairs covering the area (box "4") directly anterior to the tail.

Thus, different structures and their distribution on the cercaria body can be precisely examined under the scanning electron microscope. A comparative study using this methodology is expected to extend our taxonomical knowledge on the cercariae of the genus paragonimus which has hitherto been hindered by the limited performance of light microscopes.

Micrographs and data of this section were provided by courtesy of Dr. Yoichi Ishii, Department of Parasitology, Kyushu University School of Medicine and by permission of the Editor of the Japanese Journal of Parasitology.

Note

The primary host shell-fish, *Tricula minima* was experimentally infested with *Paragonimus sadoensis*. Cercariae were taken from the crushed shell-fish onto a glass slide and fixed with 2 per cent glutaraldehyde and coated with gold. EM: JSM-2

Reference

Ishii, Y. and I. Miyazaki: Preliminary observations on the ultrastructure of the body surface of *Paragonimus* cercaria (in Japanese). Jap. J. Parasitol. 17: 487-493 (1968).

Hockley, D. J.: Scanning electron microscopy of *Schistosoma mansoni* cercariae. J. Parasitol. 54: 1241-1243 (1968).

43. Surface Anatomy of the Flea

Fleas are blood-sucking ectoparasites of man and animals. They are characterized among other parasitic insects by their body shape, being strongly compressed laterally instead of being flattened dorsoventrally. They are very sensitive to warmth. When their host dies and its body cools, they seek fresh hosts which may not be of the same kind. Some species of fleas are known as transmitters of human diseases.

The cosmopolitan human flea named *Pulex irritans* is rather difficult to collect these days. In this chapter body structures of the familiar dog flea, *Ctenocephalides canis,* will be observed under the scanning electron microscope.

Fig. 176 (×300)

Fig. 177 (× 900)

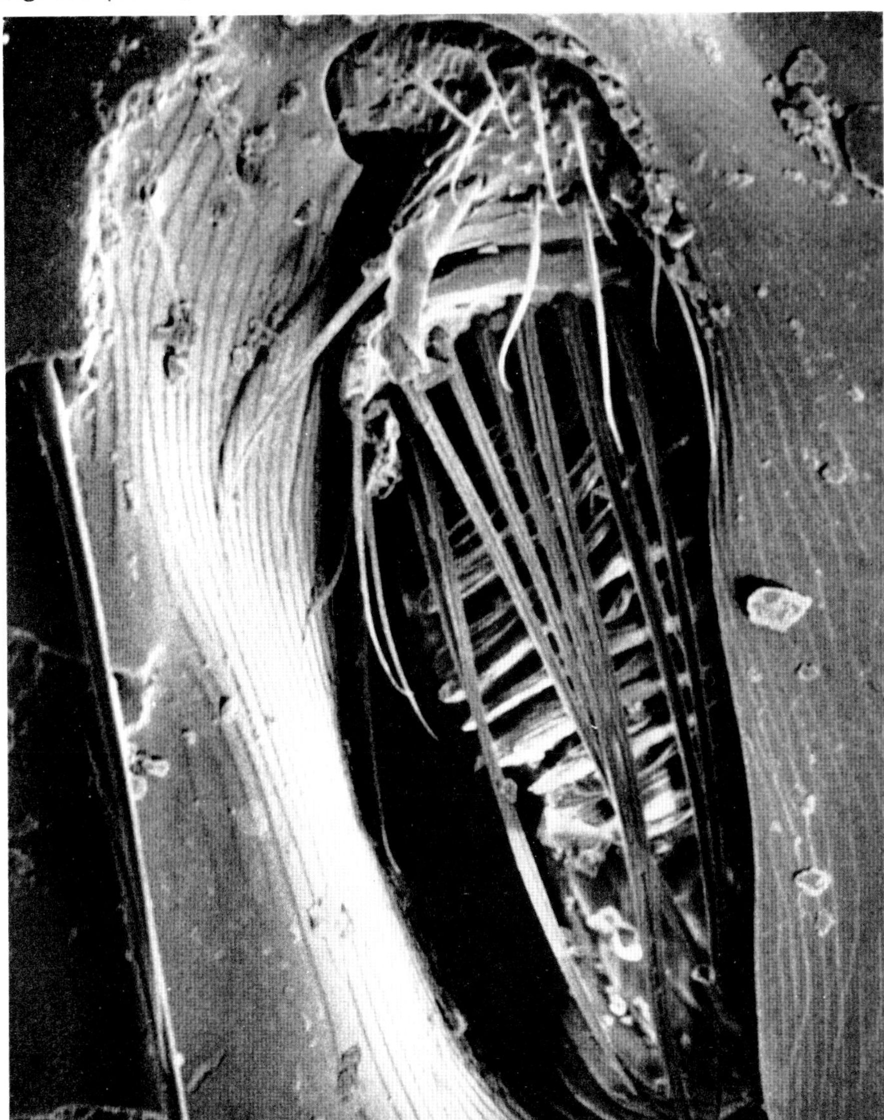

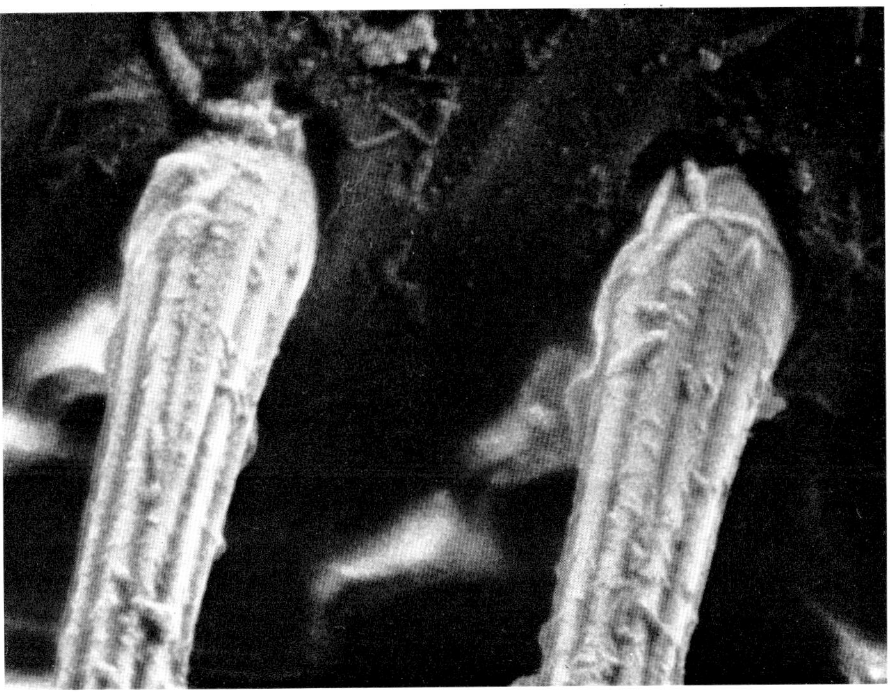

Fig. 178 (× 10,000)

Fig. 176 shows the head and thorax of a female flea from a dog. Under the characteristically domed head, there is a maxillary comb formed of eight spines. This structure, which is colored with dark pigment, is said to serve as a plow when the flea moves through obstacles. This comb is lacking in *Pulex irritans;* the uppermost, small spine of the comb is as long as the second one in the cat flea, *Ctenocephalides felis.* A complex of tentacle-like organs labeled "LP", "L" and "MP" (see below) forms the mouth parts.

The eye labeled "E" is a vestigial organ. Behind the eye is an elongated groove containing a complicated harp-like structure called antenna. This is believed to be a sensory organ but we do not know what sort of perceptory sense it has.

Fig. 177 is a closer view of the antenna. Nine or ten long bristles cover the organ and behind these a folded structure is seen. **Fig. 178** shows the root of the bristles. Each bristle, which reminds us of a Greek column, seems to be formed of several fibrous components running in parallel. It is worthy to mention that these bristles covering the antenna do not occur in *Pulex irritans.*

Fig. 179 (×900)

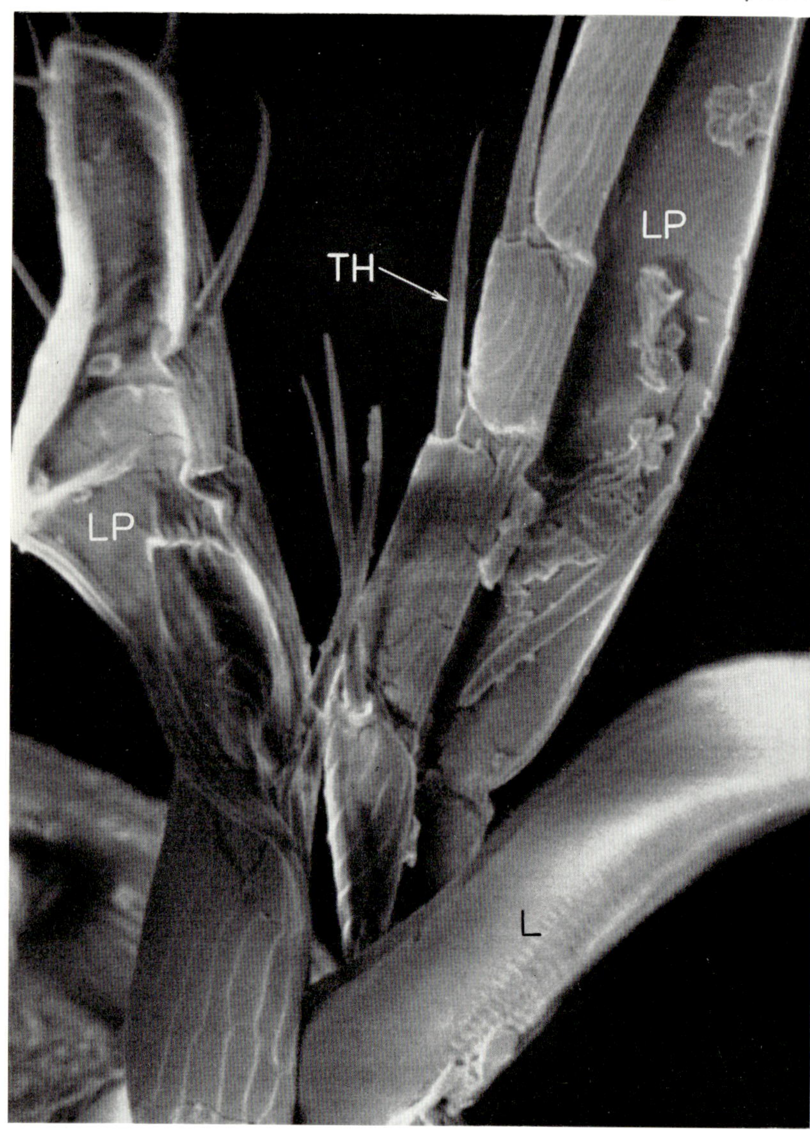

Figs. 179-183 are micrographs of the mouth parts. The needle used in blood sucking is called lacinia (labeled "L"). This is a pointed blade, finely serrated along its edges as seen in the high-power micrographs (**Figs. 180** and **183**). The lacinia is inserted into the skin of the host by a sawing up-and-down movement.

In **Figs. 179, 181** and **183** is shown a paired, segmented organ called labial palpus (LP) which helps the function of the lacinia. With its longitudinally folded structure, it can ensheath the lacinia when the latter is not in use (**Fig. 181**). The tentacles near the tip of the labial palpus are called tactile hairs and are thought to have a taste sensation.

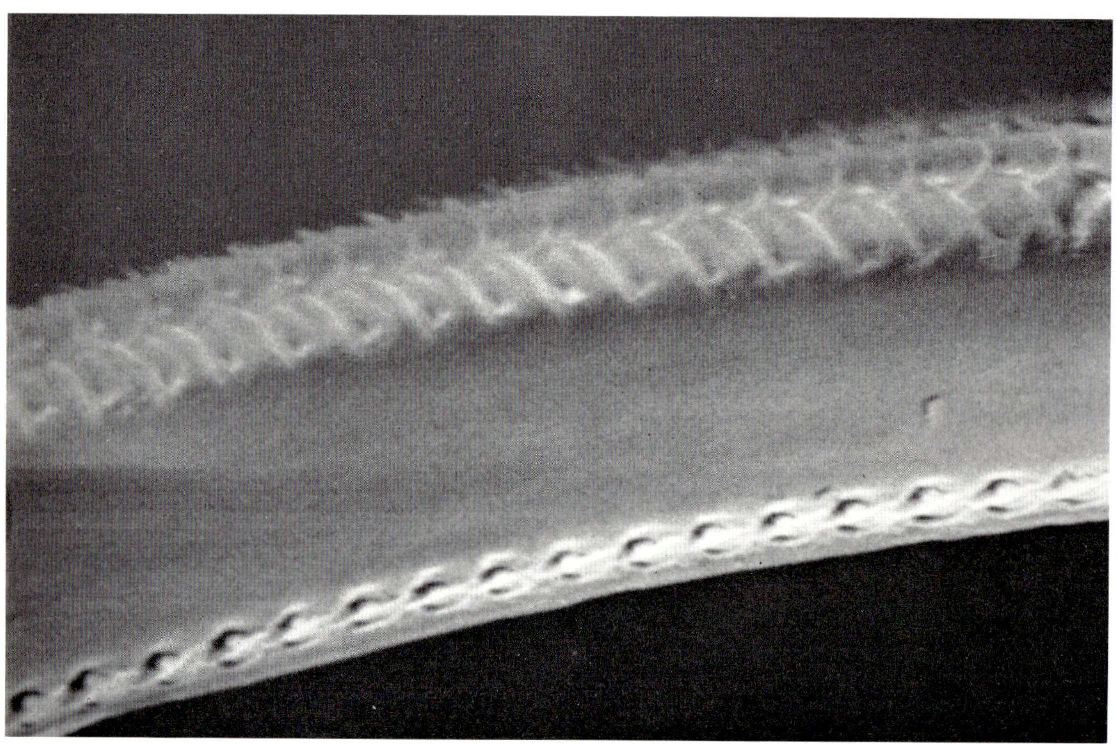

Fig. 180 (×2,700)

Fig. 181 (×720) Fig. 182 (×500)

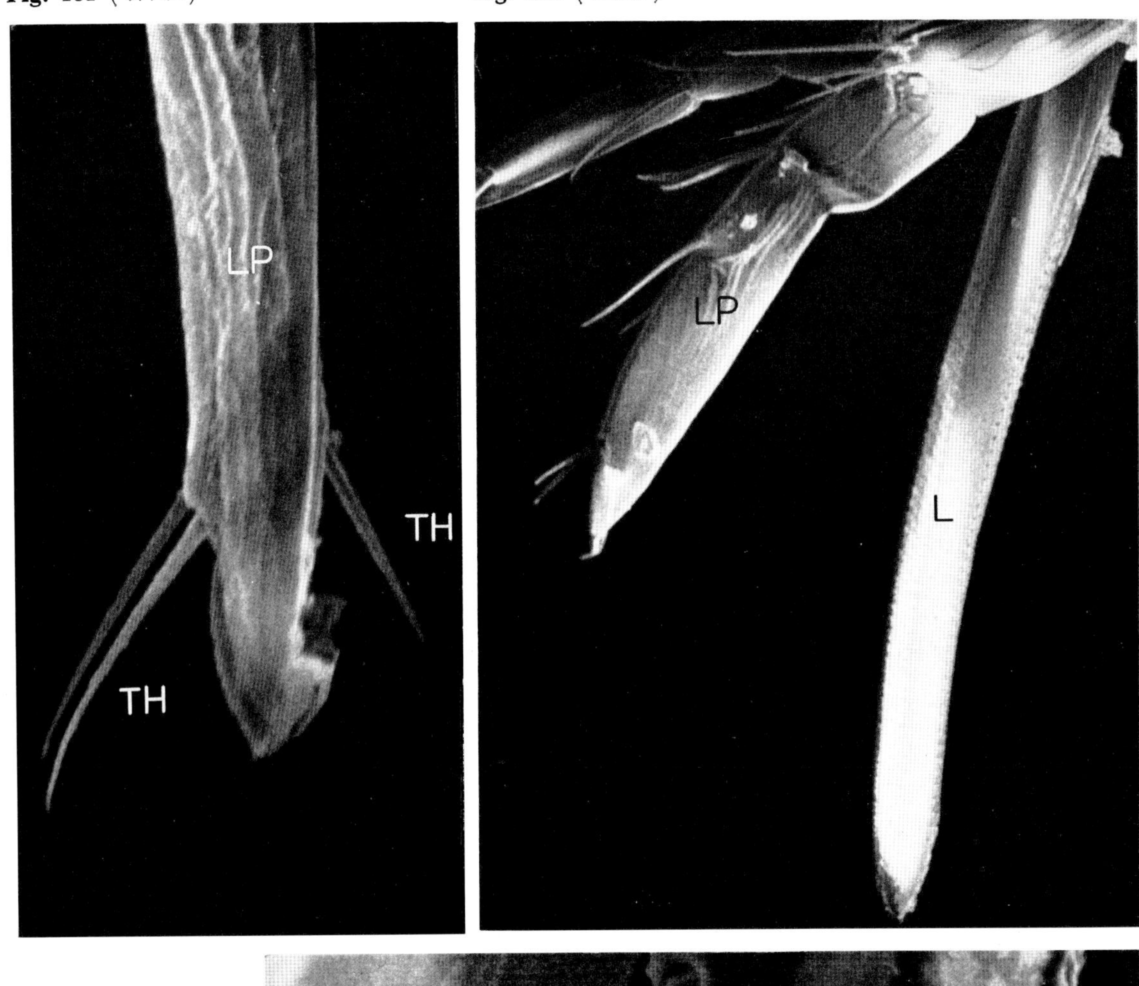

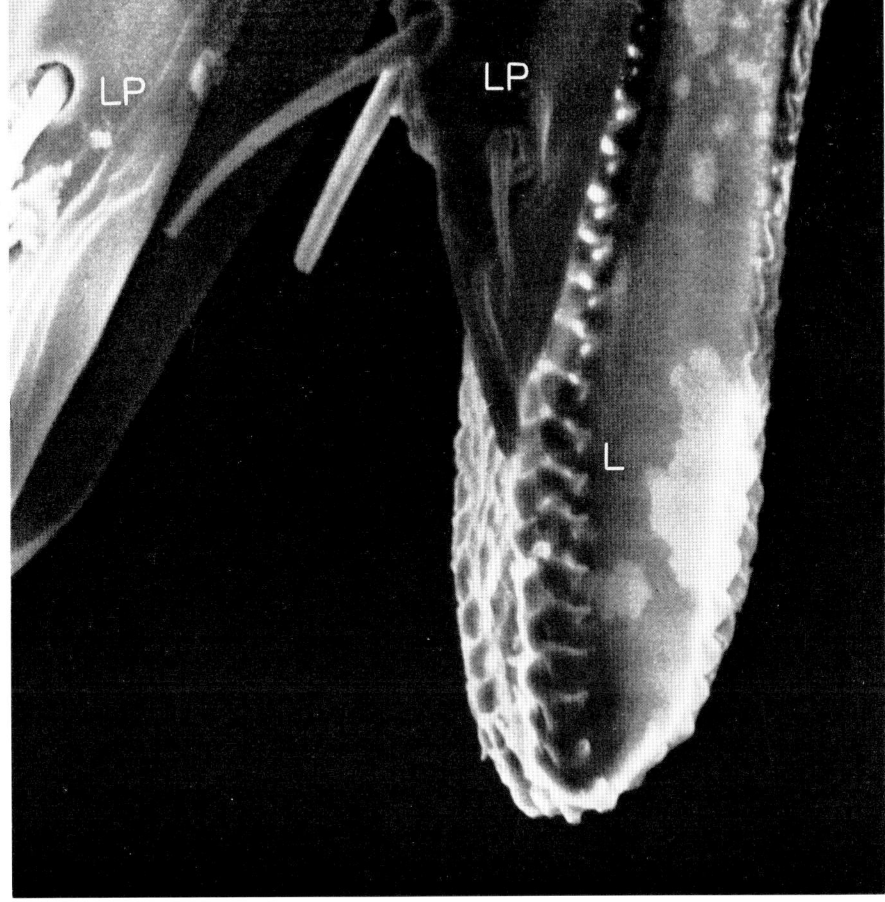

Fig. 183 (×2,500)

Fig. 184 (×4,000)

Fig. 184 shows maxillary palpus, another organ found lateral to the labial palpus. The flea, when alive, actively moves this organ. The margin of its flattened end-part is provided with motile spines called tactile cones. Some authors believe that they represent vibroreceptors.

Fig. 185 shows the hind margin of the body armed with segmented plates, called sclerites, with stout spines growing on them. In the center of this picture one may see an orifice of the Malpighian duct or trachea.

Fig. 186 is a high-power micrograph of an orifice of the trachea found at another place of the body. The wall of the duct is narrowed by a number of finger-like projections.

Fig. **185** (×1,200)

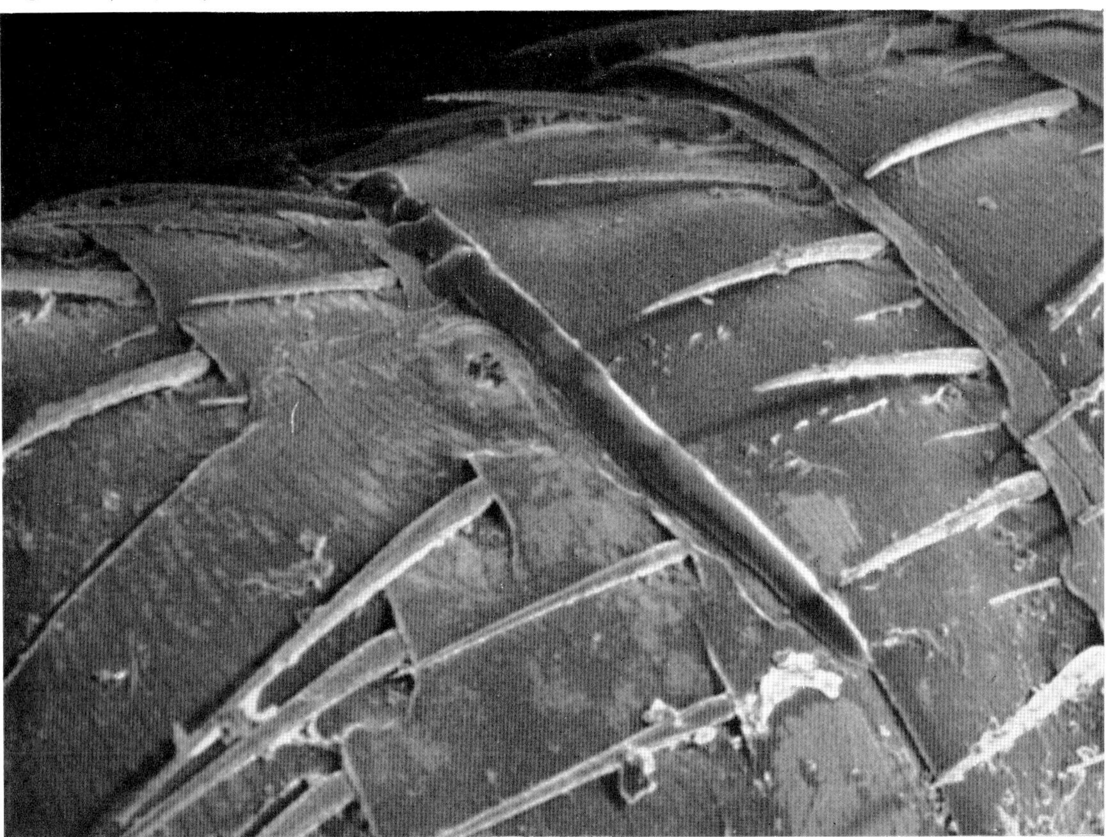

The hard cuticular armature of insects has hindered the preparation of the sections of their body for transmission electron microscopy. The figures shown in this section may indicate that the scanning electron microscope is an epoch-making weapon in the morphological study of insects, especially tiny ones such as fleas.

Note

The flea as a whole body was fixed in a 5 per cent aqueous solution of acrolein, dehydrated in acetone, dried in air and coated with carbon and gold. EM: JSM-2

Reference

HAYES, T. L., R. F. W. PEASE and A. S. CAMP: Stereoscopic scanning electron microscopy of living *Tribolium confusum*. J. Insect Physiol. 13: 1143-1145 (1967).

HINTON, H.E.: Some structures of insects as seen with the scanning electron microscope. Micron 1: 84-108(1969).

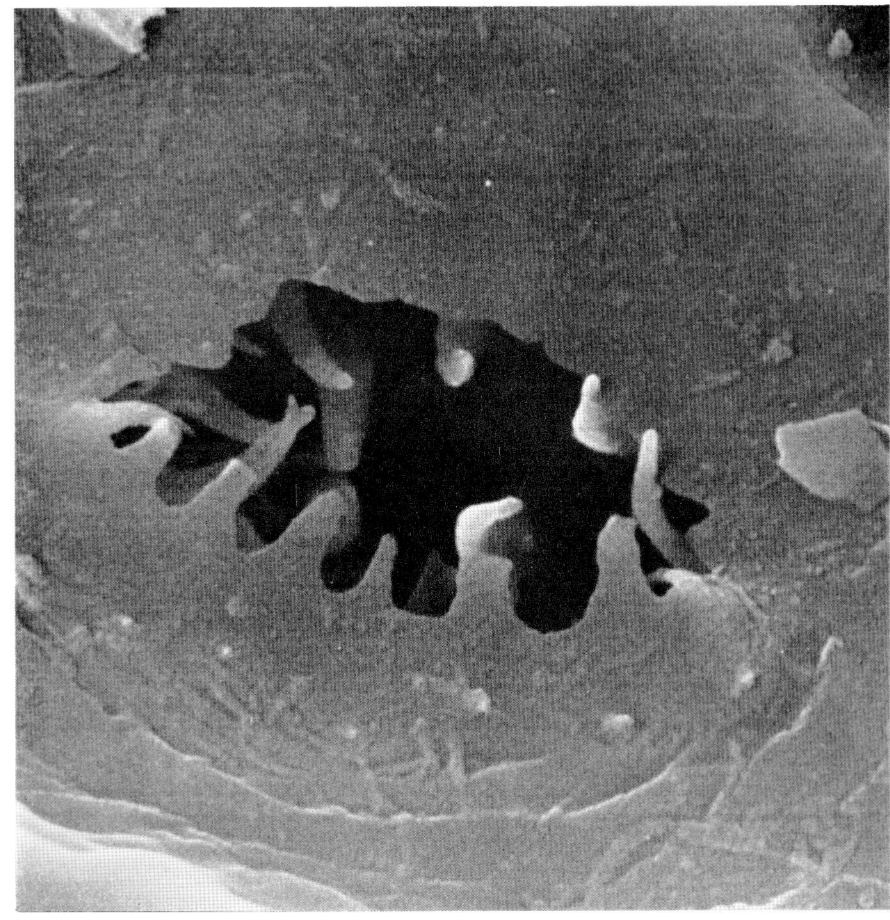

Fig. 186 (×12,000)

FUNGI AND BACTERIA

44. Aspergillus and Its Conidium Formation

Aspergillus is a genus of fungi including some common molds growing on foods. A few species are known to contain toxic substances and, when eaten, may cause severe intoxication. Molds of this group growing on stored corns and producing vicious toxin have caused a serious social problem. Furthermore, aspergillus fungi, when they occasionally become rampant within the human body, may cause aspergillosis, which has significantly increased in occurrence in recent days.

Figs. 187 and **188** are the scanning electron micrographs of conidiophores of aspergillus. Innumerable rod-like outgrowths called sterigmata are arranged radially on a globose swelling (not visible in this micrograph) at the end of the conidiophore stem. They remind us of marking pins densely stuck on a pincushion. As shown in the diagram (**Fig. 190**), conidia or asexual spores are formed by splitting off from the summit of sterigmata.

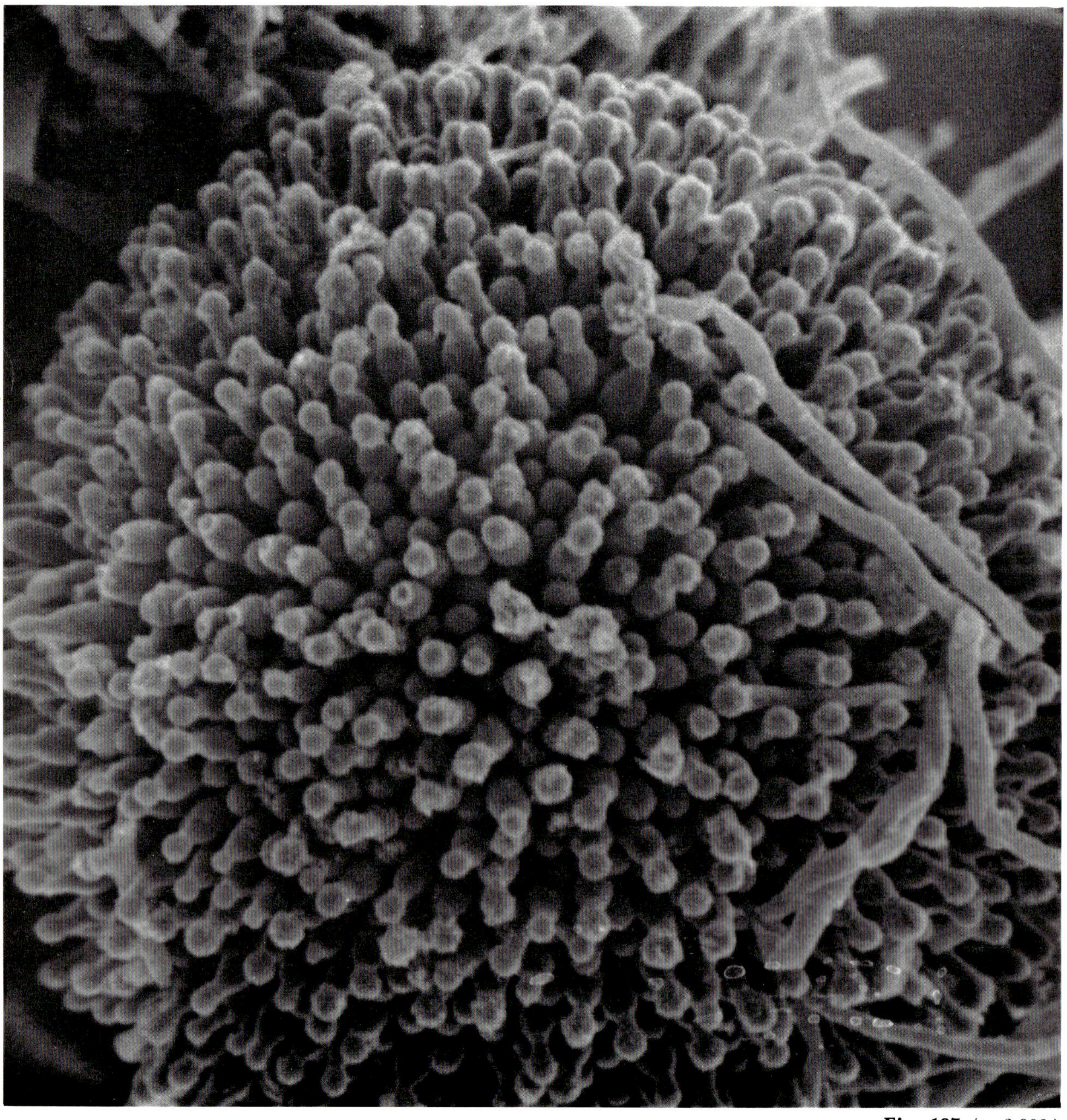

Fig. 187 (×3,000)

Fig. **188** (×4,000)

The conidiophore of **Fig. 188** shows no conidium formation as yet, but that of **Fig. 187** has round conidia on the top of the sterigmata. As a closer view (**Fig. 189**) shows, some conidia are large and nearly matured but others still tiny and seem to be just sprouting. A few sterigmata carry none of conicium buds as yet.

Fig. 191 shows a conidiophore of another species. Conidia of more or less matured types are represented by spherical bodies with characteristically undulated projections as closely shown in **Figs. 193** and **194.** The conidia in **Fig. 192** are of an immature stage and are deformed in the drying of the specimen into a pumpkin-like form.

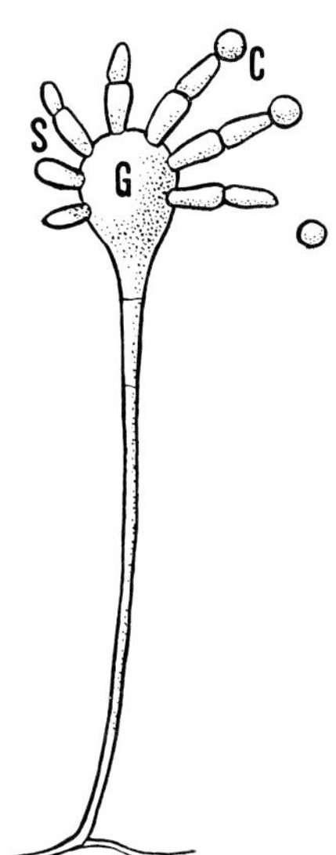

Fig. 190. Diagram showing the conidium formation of aspergillus.

G: globule at the top of the conidiophore stem. **S:** sterigmata. **C:** conidia.

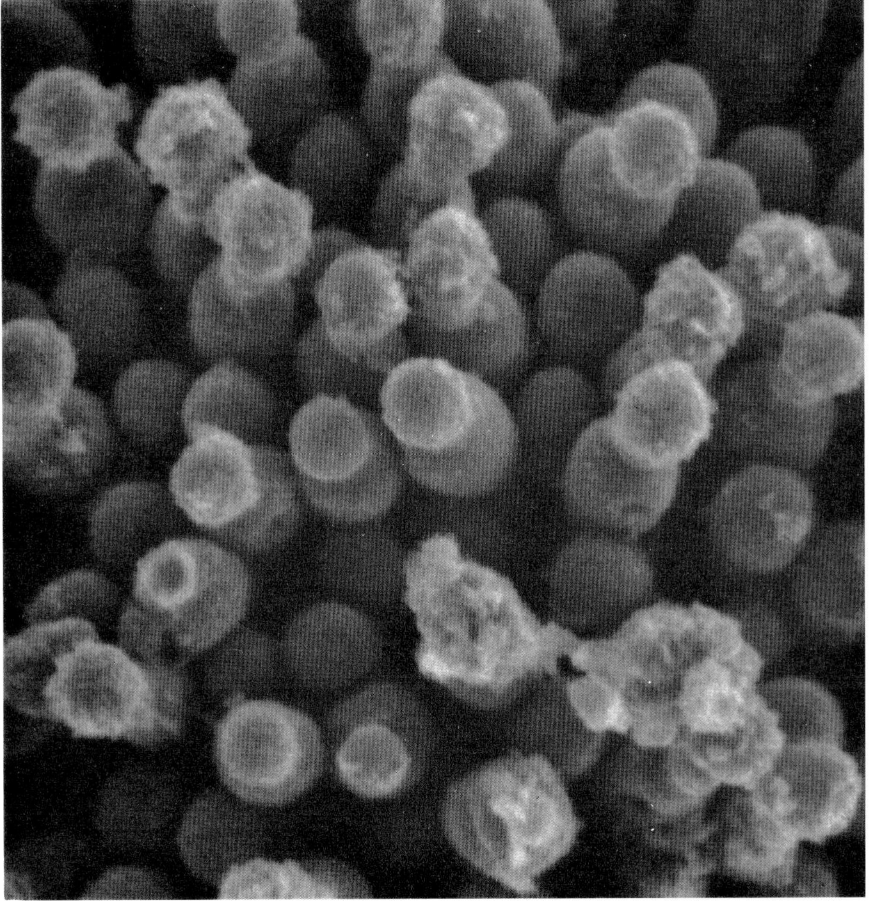

Fig. 189 (×7,200)

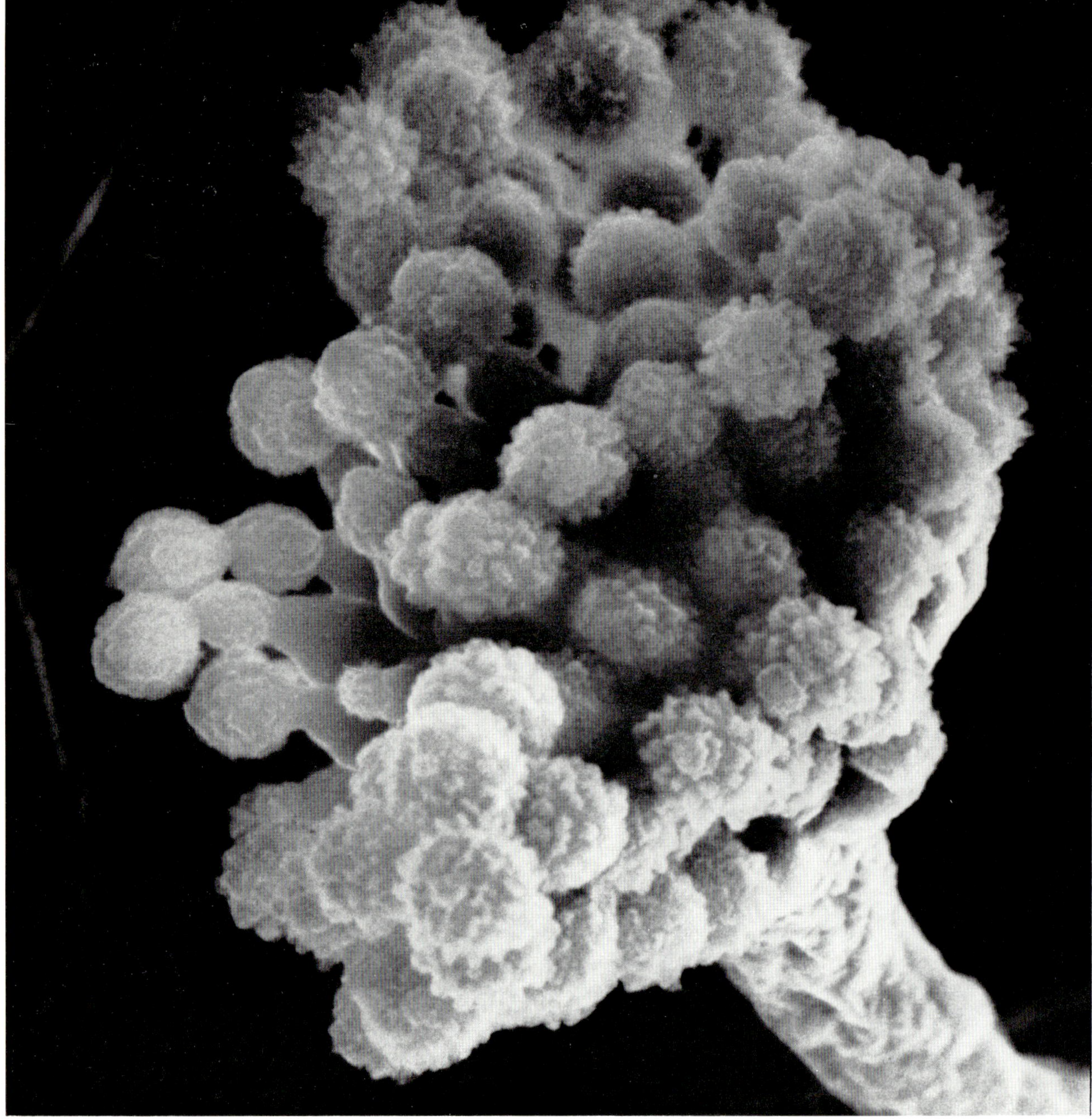

Fig. 191 (×5,100)

Some researchers propose a classification of aspergillus species based on the shapes of the conidia. The micrographs shown in this section indicate that the scanning electron microscope is the most useful weapon for studies for this purpose.

Note

Conidiophores of *Aspergillus niger* (Figs. 187, 188, 189 and 193) and *A. orizae* (Figs. 191, 192 and 194), together with a piece of their Sabouraud medium, were fixed doubly in 2.5 per cent glutaraldehyde and 2 per cent potassium permanganate, dehydrated in acetone, dried in air and coated with carbon and gold. EM: JSM-2

Fig. 192 (×12,000)

Fig. 193 (×7,200)

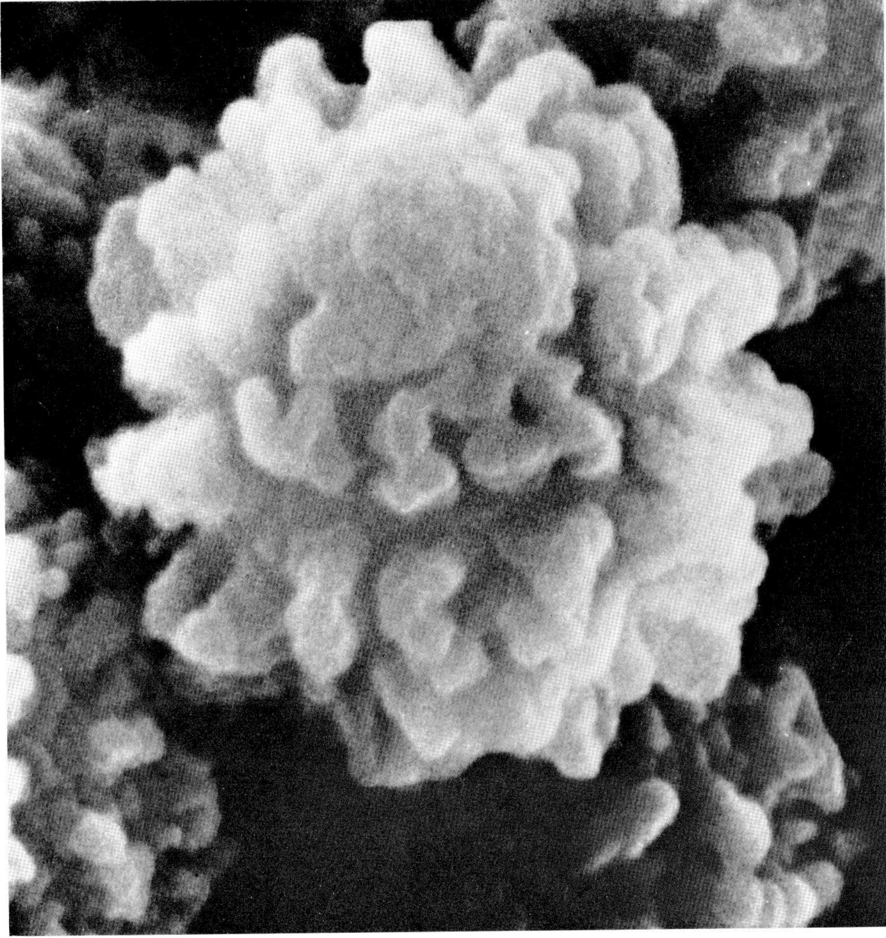

Fig. 194 (×24,000)

45. Candida albicans Microcolony on the Solid Medium

Candida, formerly called monilia, is a genus of yeast-like fungi forming smooth, white colonies (candidus: glowing white; monilia: necklace). Among large varieties of species, only *Candida albicans* causes a human infection called moniliasis. *Candida albicans* infests the mouth, bronchi, lungs, intestines and vagina as well as the skin and nails. Even sepsis, severe general infection and meningitis may occasionally be caused.

Candida albicans, when cultured on an appropriate solid medium, forms cream-like colonies composed of densely grown threads called hyphae. **Fig. 195** is a scanning electron micrograph of a part of a colony showing long, undulated filaments of hyphae.

Fig. 196 shows the center of a growing colony in which oval blast spores (yeast-like cells) are being actively formed. They are formed by budding from the hyphae which grow radially from the center of the colony.

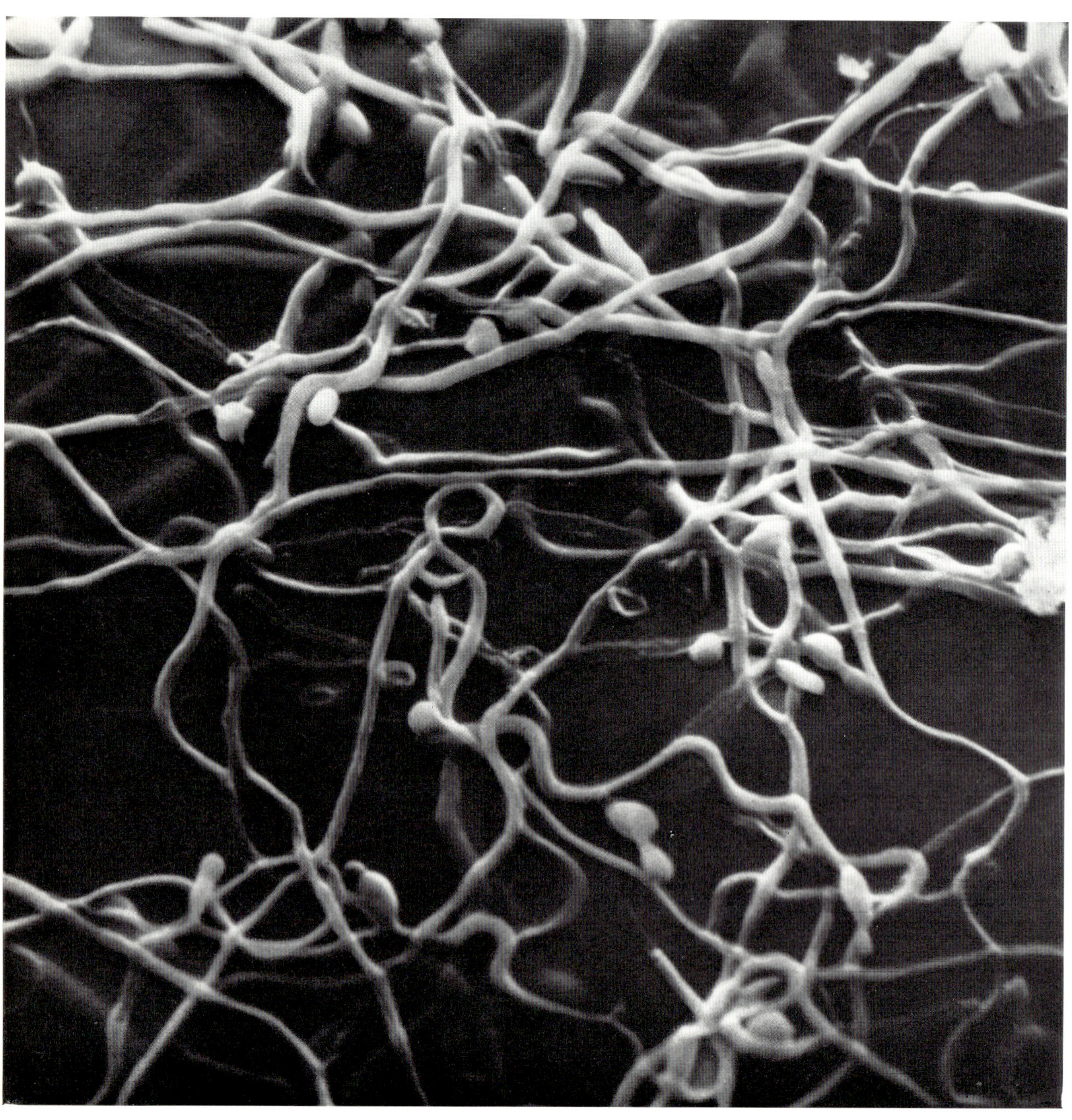

Fig. 195 (× 1,700)

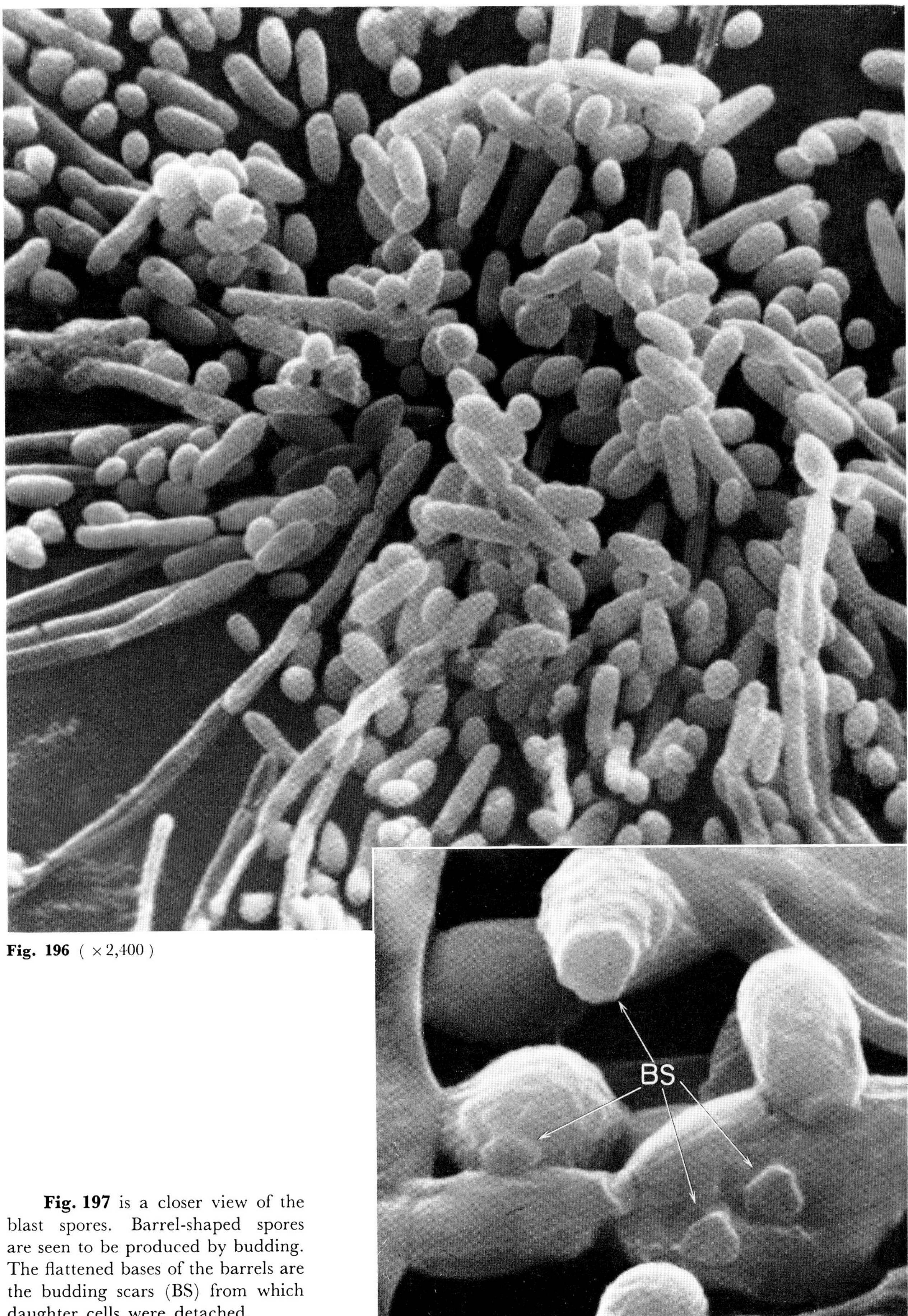

Fig. 196 (×2,400)

Fig. 197 is a closer view of the blast spores. Barrel-shaped spores are seen to be produced by budding. The flattened bases of the barrels are the budding scars (BS) from which daughter cells were detached.

Fig. 197 (×15,000)

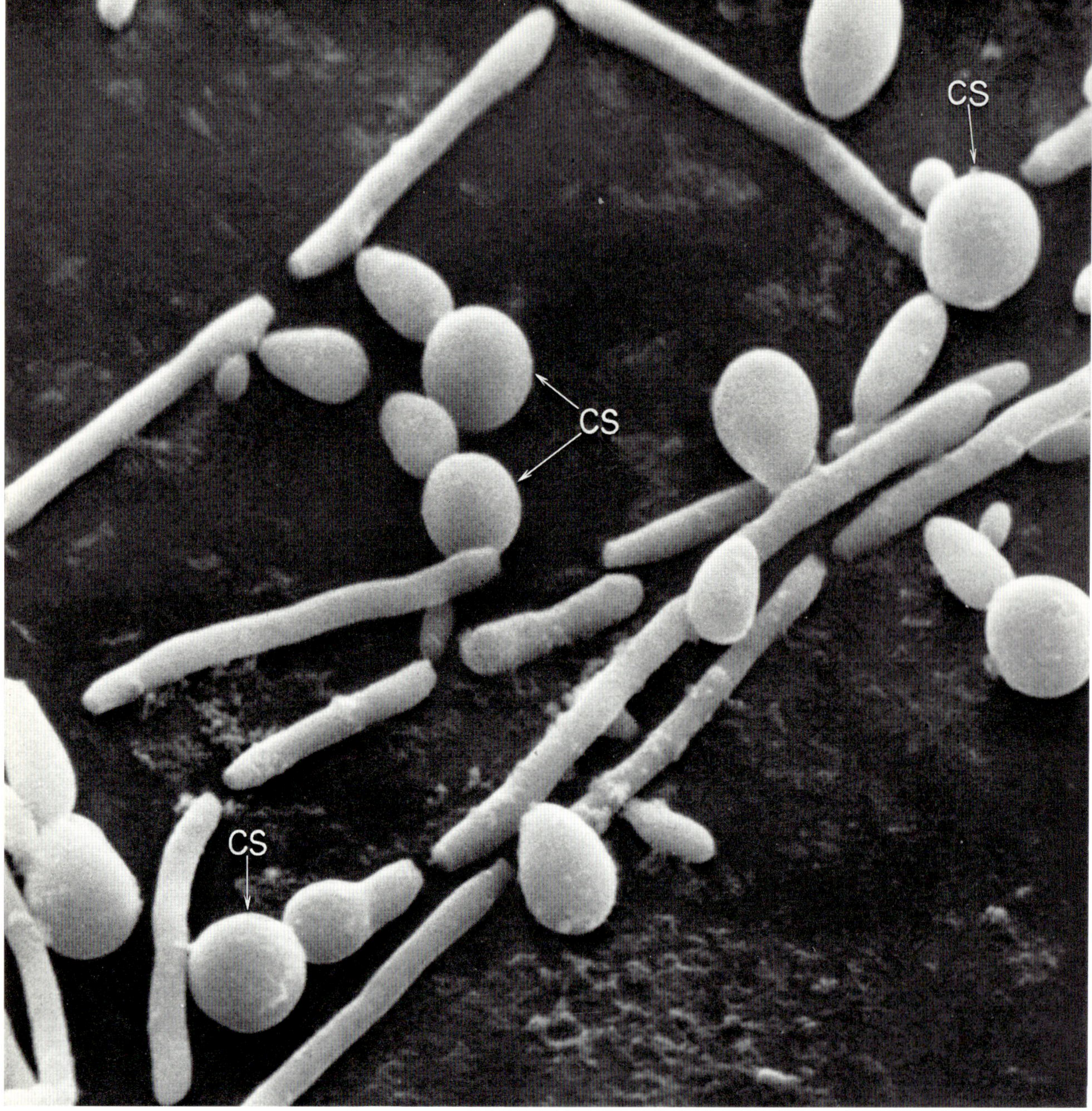

Fig. 198 (×2,800)

Fig. 198 is another aspect of the hyphae and spores. The baloon-like chlamydospores (CS) are attached to the end of the hyphae. This micrograph, which is from a specimen hardened by potassium permanganate after the aldehyde fixation, shows smoothly rounded spores and cylindrical hyphae. The other pictures of wrinkled spores and flattened hyphae indicate the shrinkage of the structures.

Figs. 199 and **200** are the side views of a culture of *Candida albicans* and this kind of photograph can be obtained only by scanning electron microscopy. An upward growth of the stem part of hyphae carrying a chlamydospore on its summit is evident.

The transmission electron microscopy of this group of fungi has been especially difficult because the spores, whose structure is critical for their species identification, are covered by a thick, hard cell wall (**Fig. 201**). Scanning electron microscopy with its three-dimensional information of fine structures is expected to contribute much to the morphology and taxonomy of monilial forms.

Fig. 199 (×600)

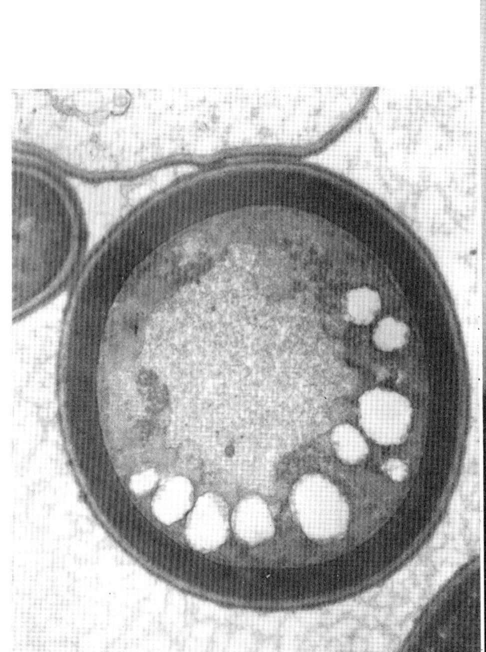

Fig. 200 (×3,600)

Fig. 201 (×8,000)

Note

Colonies of *Candida albicans* with their Sabouraud medium were fixed either singly in 2.5 per cent glutaraldehyde (Figs. 195, 196 and 197) or doubly in glutaraldehyde and 2 per cent potassium permanganate (Fig. 198). The material of Figs. 199 and 200 is not fixed. The specimens were dehydrated in acetone, dried in air and coated with carbon and gold. EM: JSM-2. Fig. 201 is a transmission electron micrograph.

46. Bacterial Flora on the Tongue and in Pus

Scanning electron microscopy enables us to observe the bacterial flora as they grow on the surface and in the tissues of the body.

Fig. 202 shows the surface of a part of a human tongue excised in surgery. The patient suffered from glossitis dissecans, a chronic disease in which the tongue is deeply furrowed. As seen in this micrograph, the epithelial cells of the tongue are densely covered with bacterial colonies. Numerous round cocci, mainly about 1μ in diameter, and a few rod-like bacilli are distinguished. Some cocci show a cleavage line which is a constriction halving the cell. The large white body on the left side is a red blood cell which attached to the specimen in the operation.

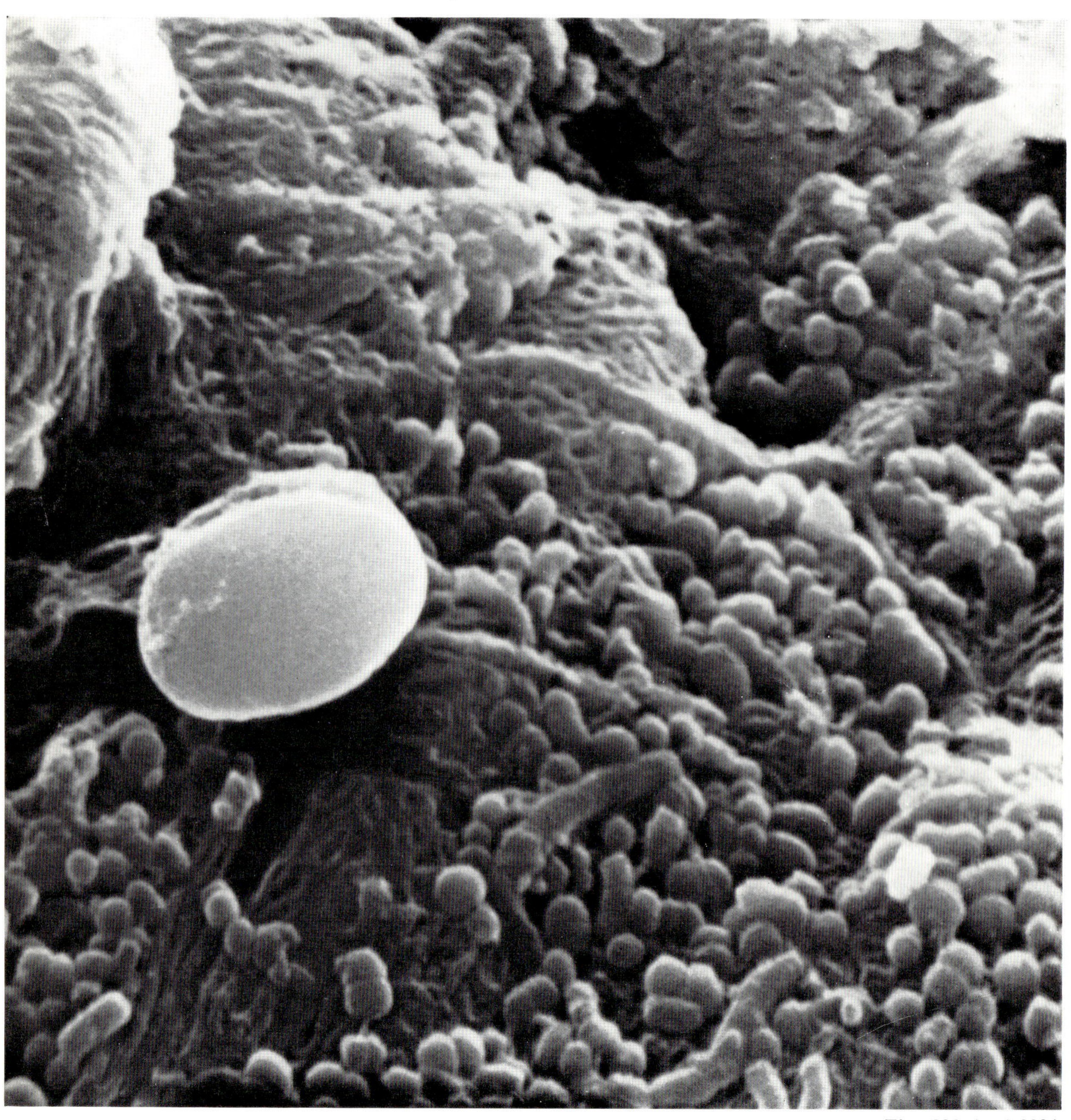

Fig. 202 (×9,000)

Fig. **203** (× 15,000)

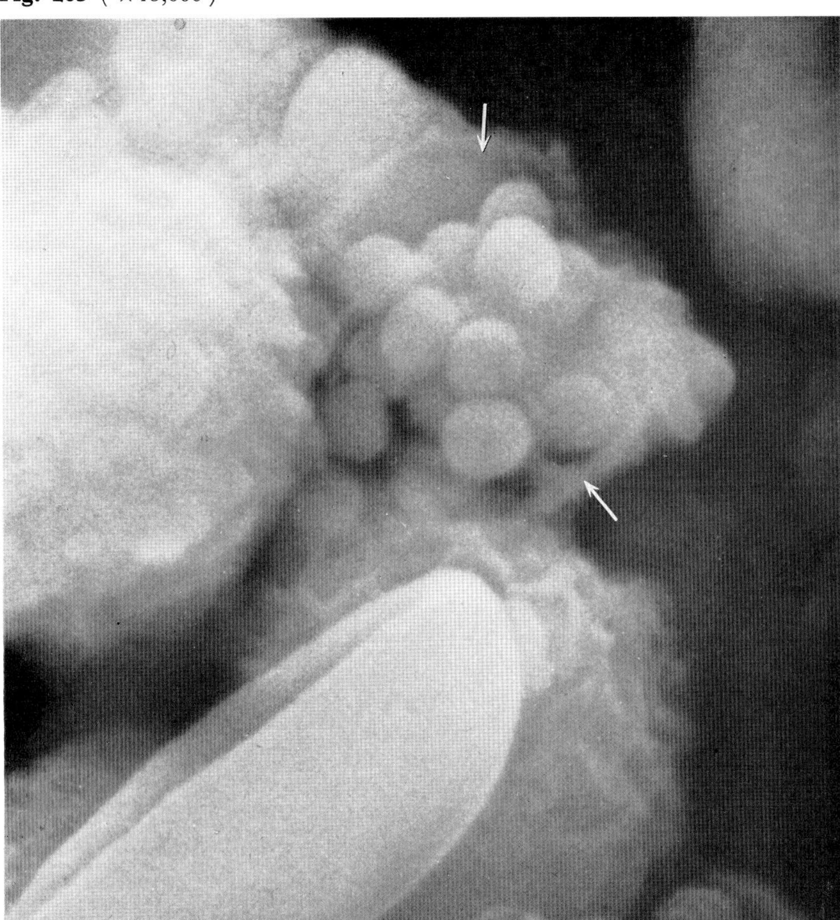

Fig. 204 (× 12,000)

Figs. 203 and **204** are the micrographs of the cells and bacteria in pus. An inflammation on the fore-arm of one of the authors (J.T.) was left, for the purpose of study, without any treatment. When an abscess developed, the pus was taken and observed under the scanning electron microscope.

A bunch of grapes shown in **Fig. 203** corresponds to staphylococci which have caused this inflammation. Covering the bunch is a membrane resembling a broken bag (arrows). This may be parts of a white blood cell which possibly phagocytosed the cocci and then was destroyed by them. A few white blood cells of round form and a red cell (below) are seen in this picture.

Fig. 204 indicates white blood cells which represent the substance of pus. Vesicle-like spots are seen in one of the cells and correspond in size to the cocci, many of which are incorporated by the cells.

Fig. 202 was provided by courtesy of Dr. TAKUO JOZAKI of the Yahata-Seitetsu Hospital.

Note

The pieces of the tongue were simply washed, fixed in 10 per cent formalin, dehydrated in acetone, dried in air and coated with gold. The pus was fixed in 2.5 per cent glutaraldehyde and further treated as in the case of blood (p. 75).
EM: JSM-2

47. Staphylococci in Division

Staphylococci are a group of microbes widely distributed in nature. In human and animal bodies they are found on the skin, in the nasal and oral cavities and in the digestive tract. They occur abundantly in suppurative foci. Especially vicious in pathogenic nature is the species called *Staphylococcus aureus* which is common in the human upper respiratory tract. The cocci growing in the nasal vestibulum are directly involved in infection. This group of microbes has recently attracted attention with regard to hospital infection and the occurrence of strains tolerant to antibiotics.

The name of staphylococcus comes from its grape-like appearance under the light microscope (staphyle: a bunch of grapes). As shown in **Fig. 205,** a microcolony of *Staphylococcus aureus* on a solid medium of agar seen under the scanning electron microscope resembles a bunch of grapes lying on the table.

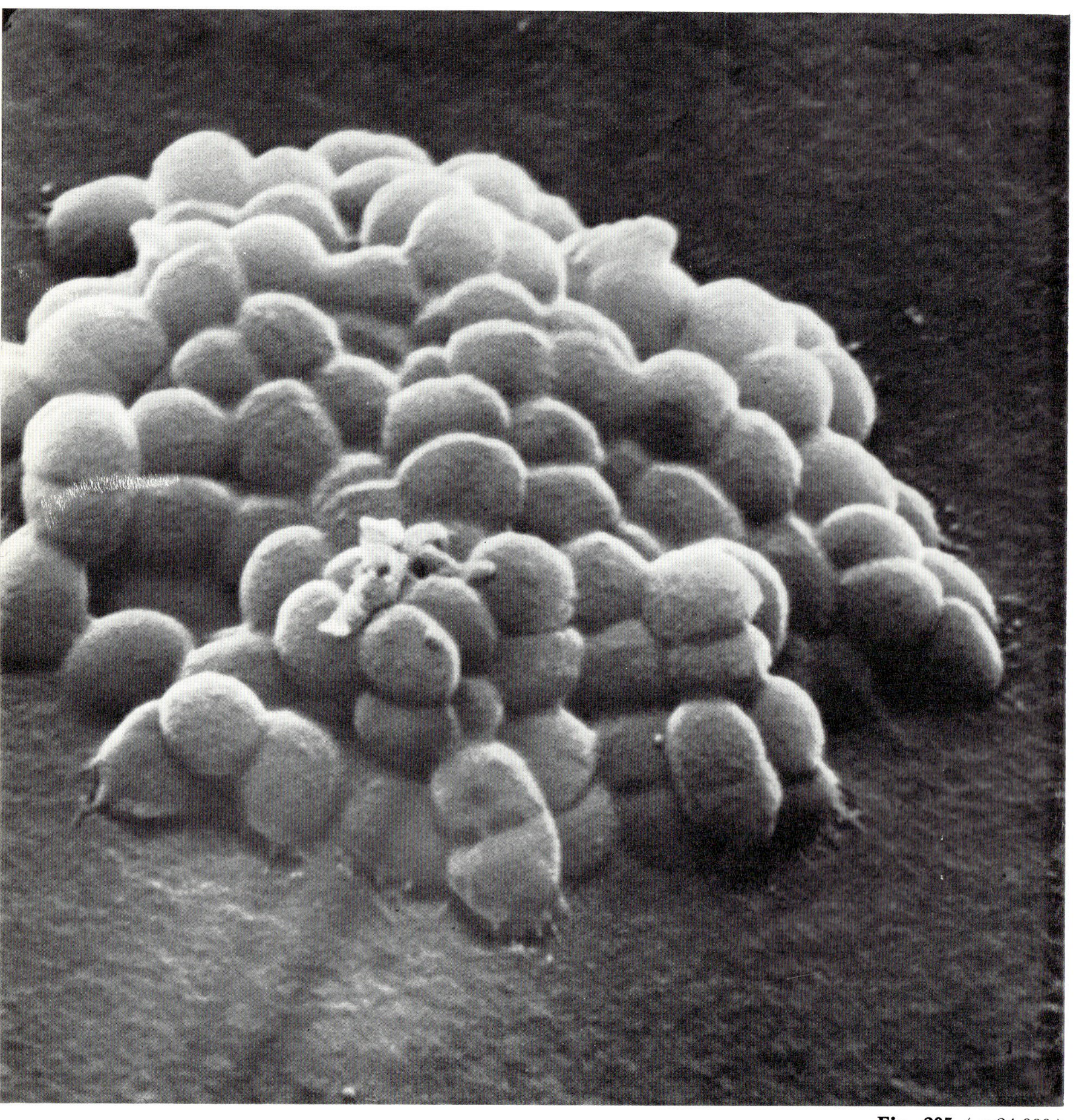

Fig. 205 (×24,000)

Fig. 206 (×36,000)

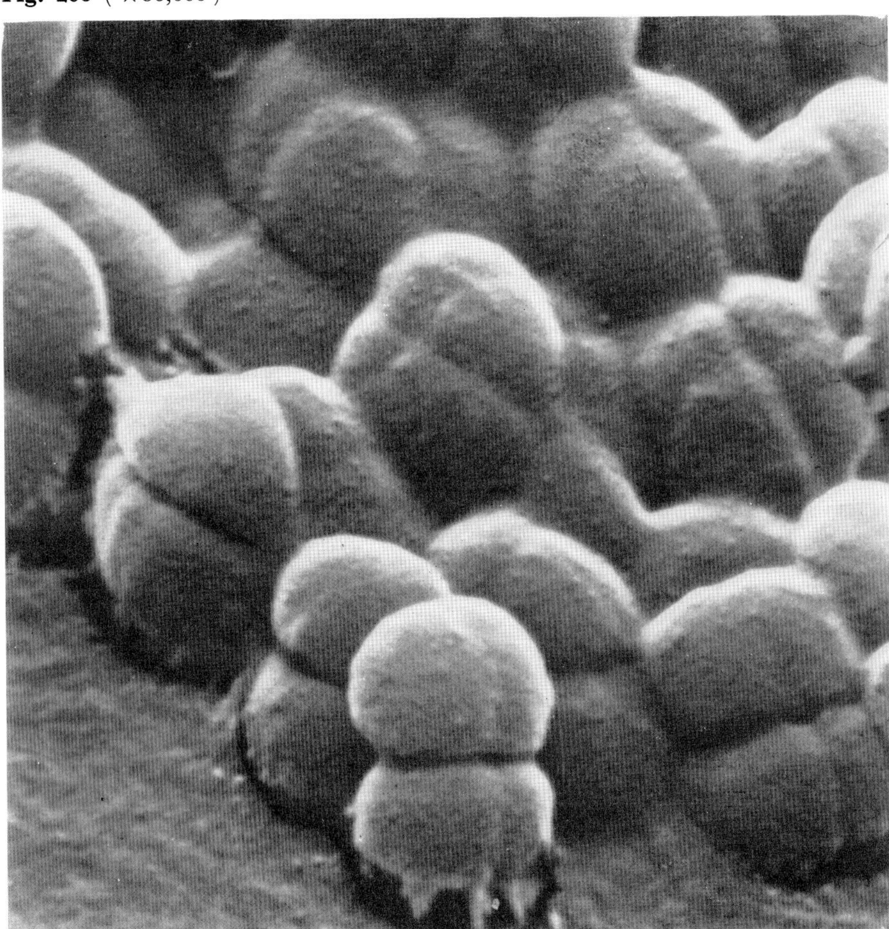

Fig. 206 shows the surface structure of individual bacterial cells in a closer view. A single line or double, crisscross lines of cleavage are seen on the surface of most cells. This indicates that an active proliferation of bacterial cells is taking place in this colony. The divided cells are pushed into an accumulation of polyhedral forms.

Beautiful spherical forms, however, can be obtained when the same material cultured in broth is harvested and rinsed. **Fig. 207** is a micrograph from this sampling. Some cells having a cleft look like peaches.

Note

Agar medium carrying microcolonies of *Staphylococcus aureus* was cut out under the binocular. After fixation in 2.5 per cent glutaraldehyde and then in 2 per cent osmium tetroxide, the specimens were dehydrated in acetone, dried in air and coated with carbon and gold. The specimen of Fig. 207 was made by harvesting the cocci cultured in broth. After being rinsed in water, treated in the same fixatives and dehydrated in acetone, they were spread on glass slides and coated with carbon and gold. EM: JSM-2

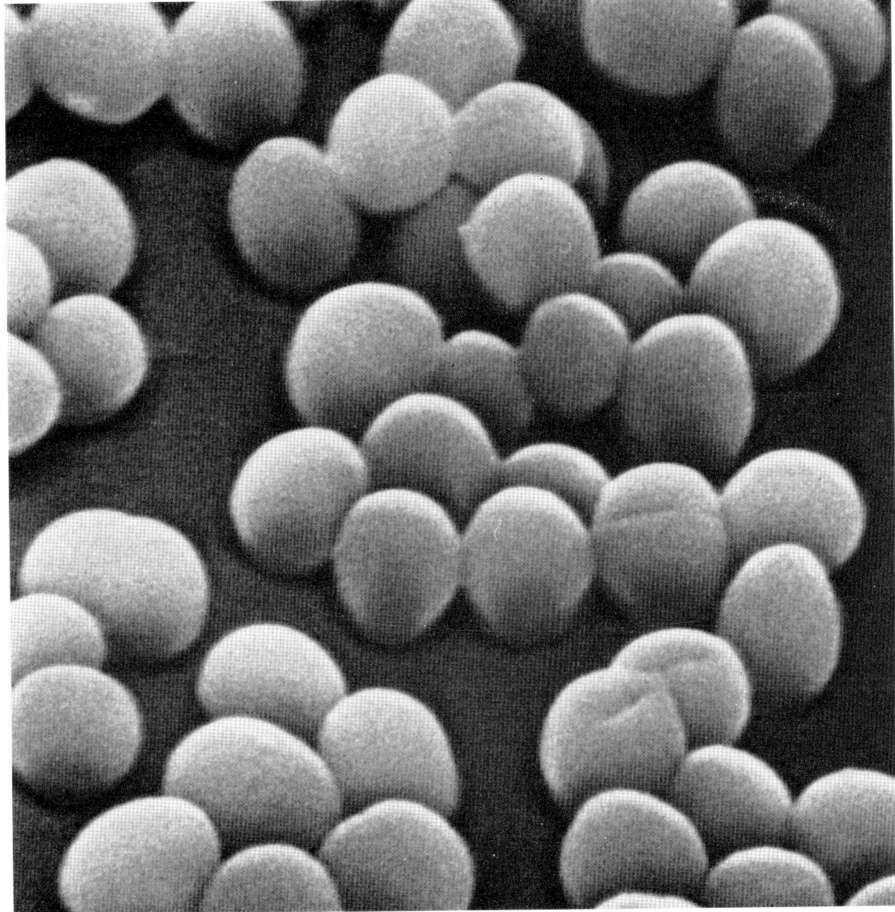

Fig. 207 (×23,000)

INDEX